GW00994694

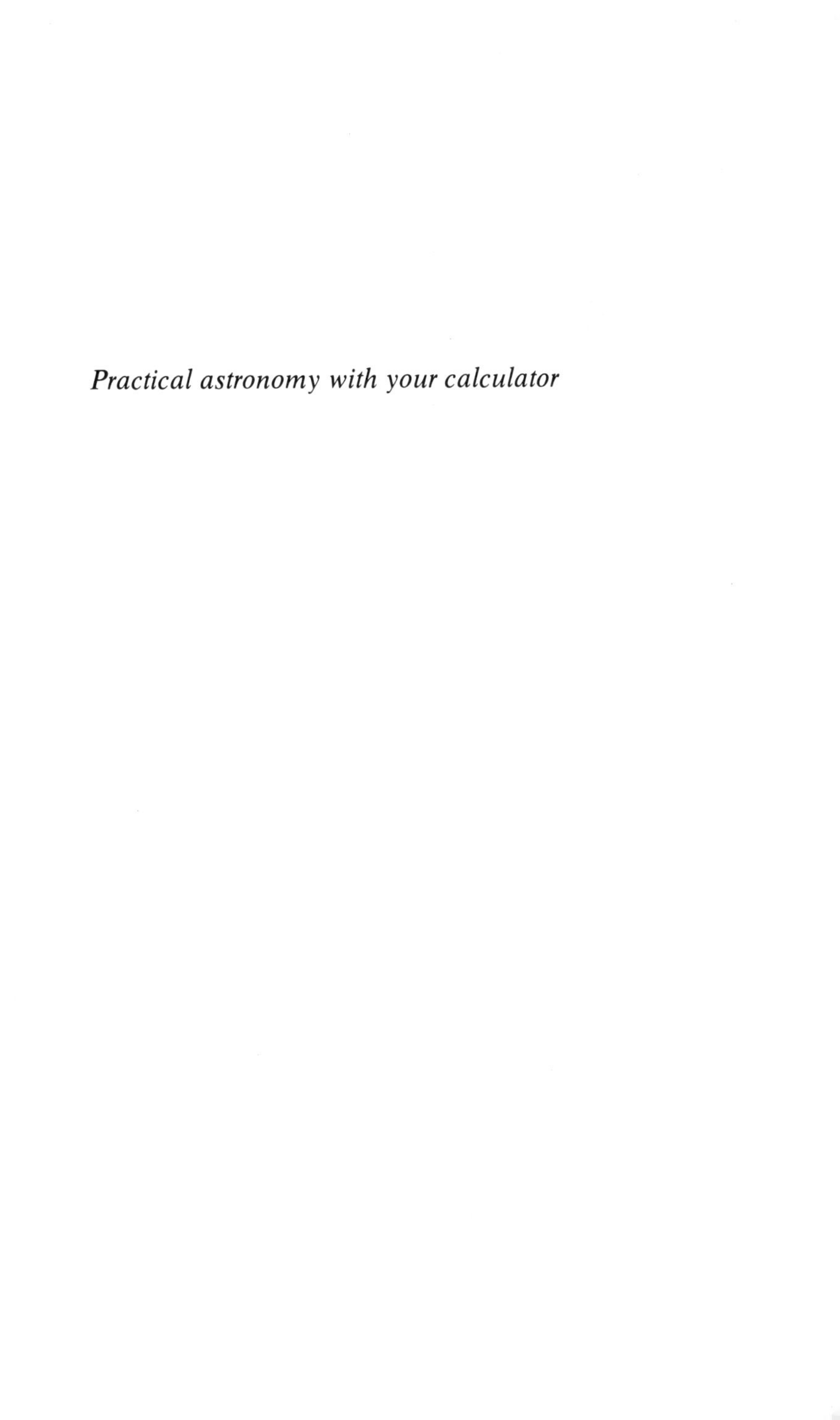

Practical astronomy with your calculator

Practical astronomy with your calculator

PETER DUFFETT-SMITH

University Demonstrator
Mullard Radio Astronomy Observatory, Cavendish Laboratory, and
Fellow of Downing College, Cambridge

SECOND EDITION

CAMBRIDGE UNIVERSITY PRESS

Cambridge

London New York New Rochelle

Melbourne Sydney

Published by the Press Syndicate of the University of Cambridge
The Pitt Building, Trumpington Street, Cambridge CB2 1RP
32 East 57th Street, New York, NY 10022, USA
296 Beaconsfield Parade, Middle Park, Melbourne 3206, Australia

First published 1979
Reprinted 1980
Second edition 1981
Reprinted 1982 (twice)

Printed in the United States of America

Library of Congress catalogue card number: 81-6191

British Library cataloguing in publication data

Duffett Smith, Peter
Practical astronomy with your calculator. – 2nd ed.

1. Astronomy – Mathematics
2. Calculating – machines
I. Title

522'.028'54 QB47

ISBN 0 521 24059 X hard covers
ISBN 0 521 28411 2 paperback
(First edition ISBN 0 521 22761 5 hard covers
0 521 29636 6 paperback)

To FD
and Pickle

Contents

Preface to the second edition

Practical astronomy with your calculator has been written for those who, either for practical purposes or simply because they enjoy making predictions, wish to calculate the positions and visual aspects of the major heavenly bodies and important phenomena such as eclipses. I have tried to cut a path through the complexities and difficult concepts of rigorous mathematics, taking account only of those factors which are essential to each calculation and ignoring the corrections for this and that, necessary for the very precise predictions of astronomical phenomena. My simple methods are usually sufficiently accurate for all but the most exacting amateur astronomer, but they should not be used for navigational purposes. For example, the times of sunrise and sunset can be determined to within one minute and the position of the Moon to within one fifth of a degree.

In the second edition, I have included much more material in response to letters and requests from readers of the first edition. I have also corrected the errors which have come to my notice (but no doubt have committed different ones here) and have brought the material up to date by assuming the epoch 1980 January 0.0 as the starting point for orbital calculations.

I am most grateful to all those kind people who have taken the trouble to write to me with their suggestions, criticisms and corrections, in particular to Mr S. Hatch, Mr E. R. Wood and Mr S. J. Garvey (who supplied the nomogram for the solution of Kepler's equation). I would also like to thank and acknowledge those authors whose books I have read and whose ideas

I have cribbed, mentioning particularly Jean Meeus (*Astronomical Formulae for Calculators*) and W. Schroeder (*Practical Astronomy*).

My thanks are also due to Dr Anthony Winter, who suggested I write the book in the first place, to Mrs Dunn whose careful typing makes her the sweetheart of any author, to Dr Guy Pooley who read the manuscript and made many helpful suggestions, and to Dr Simon Mitton, Senior Editor at Cambridge University Press, for taking so much trouble over the book.

The calculations given in the examples were made with a Hewlett Packard HP67 pocket calculator.

P J D S
Downing College
Cambridge
September 1980

About this book and how to use it

How many times have you said to yourself 'I wonder whether I can see Mercury this month?' or 'What will be the phase of the Moon next Tuesday?' or even 'Will I be able to see the eclipse in Boston?' Perhaps you could turn to your daily newspaper to find the answer or go down to the local library to consult the *Astronomical Ephemeris*. You may even have an astronomical journal containing the required information. But you would not, I suspect, think of sitting down and calculating it. Yet in this modern age of hand-held micro-miniaturised integrated electronics, calculations are *easy*. Even if you have no qualifications in mathematics at all, you can calculate the answer to almost every astronomical question you are likely to ask. All you need is this book, a calculator, a piece of paper, a pencil and a ruler; by following the simple step-by-step instructions contained in the section appropriate to your question you will find the answer.

Your calculator does not have to be a very sophisticated device costing a great deal of money; on the other hand it should be a little better than a basic four-function machine. At a minimum it should be able to calculate the trigonometric functions sine, cosine and tangent (and their inverses) for any angle expressed in degrees or radians. (Beware of those calculators which only work over a limited range of angles.) It should also be able to find the square root and the logarithm of a number. Features other than these are not essential but naturally can make calculations easier. For example, a number of separately-addressable memories in which you can store intermediate results would be useful. If you have a programmable calculator, you can write programs to carry out most of the calculations automatically with a subsequent saving in time and effort; I have done this with my own calculator.

There is now a very wide range of pocket calculators available on the market, and the prices seem to continue to decrease. As a rough guide, you ought to be able to buy a sufficiently good machine for about twice the cost of the hardcover version of this book (five times the paperback) though of course you can spend very much more and save yourself much effort in later calculations on the more powerful machines. When choosing a calculator, do not be led astray by arguments over whether 'Reverse Polish Notation' (RPN) or 'Algebraic Notation' (AN) is the better system. Each has its advantages and the same complexity of calculation may be made using either. It is important for you to read the instructions carefully and to get to know your new calculator thoroughly, whether it uses RPN or AN, so that you can make calculations quickly and accurately. Make sure that you like the 'feel' of the keyboard and that pressing a key just once results in just one digit appearing in the display, rather than a whole string as the calculator responds to a bouncy and badly-made key. Finally, be on the lookout for special functions which may help you; for example, a key which converts a time or angle expressed as hours, minutes and seconds into decimal hours, a key which takes any angle (positive or negative) and returns its value in the range 0° to 360°, and a key which converts between rectangular and polar coordinates. (This last may be used very effectively to overcome the ambiguity of 180° introduced on taking inverse tan.)

When you go through the worked examples given with each calculation, do not be alarmed if your figures do not agree with mine in the last decimal place. The reason for this discrepancy may simply be that the internal accuracy of your calculator, that is, the number of figures with which it works, may not be quite as good as that of the calculator I used. Provided that your calculator has seven or eight digit accuracy, you should find very little error in the final result. A word here about microprocessors: these devices, like all computers, require a *language* by means of which the user can program the machine. Many languages, such as BASIC, which are suitable for the relatively small capacity of a home-computer system, use four bytes to represent decimal numbers in binary form and therefore have a precision of only six or perhaps seven significant figures. Some care must be used to ensure that rounding errors do not become significant.

Having gathered together your writing material, calculator and book, how do you proceed? Let us take as an example the problem of finding the time of moonrise. Turn to the index and look up 'moonrise'; you are directed to section 66 where you will find a paragraph or so of explanation and a list of instructions, together with a worked example. You need not even read the paragraphs of explanation to carry out the calculation! In any case, I have kept the explanations brief and I have not attempted to derive the formulae used; if you wish to see where they came from you should read one of the standard texts on spherical astronomy such as the excellent *Textbook on Spherical Astronomy* by W. M. Smart (Cambridge University Press, 1977). As you work through each instruction, write down its number and the result in an orderly fashion. This will help you to keep track of where you are, and to check your calculations later. If you are not methodical you may find it impossible to get the right answer.

Many calculations require you to turn back and forth between different sections. For example, instruction 2 of 'moonrise' directs you to section 62. Make the calculations in that section and then turn back to carry on with the next instruction, number 3. You'll find it useful to keep several slips of paper handy as bookmarks.

This book is not intended to replace the *Astronomical Ephemeris.* One can hardly compete with the sophisticated computers used in the yearly task of compiling that reference work. However, the accuracy of the methods given here is good enough to meet most circumstances, simplicity being more important than precision in the *n*th decimal place. If you own a home computer you can make use of this book to write programs to produce television displays of the evolving Solar System with an accuracy better than the resolution of the screen. But those of us just with simple pocket calculators can find great satisfaction in simply being able to work out the stars for ourselves and to predict astronomical events with almost magical precision.

Time

Astronomers have always been concerned with time and its measurements. If you read any astronomical text on the subject you are sure to be bewildered by the seemingly endless range of times and their definitions. There's universal time and Greenwich mean time, apparent sidereal time and mean sidereal time, ephemeris time, local time and mean solar time, to name but a few. Then there's the sidereal year, the tropical year, the Besselian year and the anomalistic year. And be quite clear about the distinction between the Julian and Gregorian calendars! (See the Glossary for the definitions of these terms.)

All these terms are necessary and have precise definitions. Happily, however, we need concern ourselves with but a few of them as the distinctions between them become apparent only when very high precision is required.

1 Calendars

A calendar helps us to keep track of time by dividing the year into months, weeks and days. Very roughly speaking, one month is the time taken by the Moon to complete one circuit of its orbit around the Earth, during which time it displays four phases, or quarters, of one week each, and a year is the time taken for the Earth to complete one circuit of its orbit around the Sun. By common consent we adopt the convention that there are seven days in each week, between 28 and 31 days in each month (see Table 1) and 12 months in each year. By knowing the day number and name of the month we are able to refer precisely to any day of the year.

The problem with this method of accounting the days in the year lies in the fact that, whereas there is always a whole number of days in the civil year, the Earth takes 365.2422 days to complete one circuit of its orbit around the Sun. (This is the *tropical year*; see the Glossary for its definition.) If we were to take no notice of this fact and adopt 365 days for every year, then the Earth would get progressively more out of step with our system at a rate of 0.2422 days per year. After 100 years the discrepancy would be 24 days; after 1500 years the seasons would have been reversed so that summer in the northern hemisphere would be in December. Clearly, this system would have great disadvantages.

Julius Caesar made an attempt to put matters right by adopting the convention that three consecutive years have 365 days followed by a *leap year* of 366 days, the extra day being added to February whenever the year number is divisible by 4. On average, his civil year has 365.25 days in it, a fair approximation to the tropical year of 365.2422 days. Indeed, after 100 years the error is less than one day. This is the *Julian calendar* and it worked very well for many centuries until, by 1582, there was again an appreciable discrepancy between the seasons and the date. Pope Gregory then improved on the system by abolishing the days October 5th to October 14th 1582 inclusive so as to bring the civil and tropical years back into line, and by missing out three days every four centuries. In

Table 1

January	31	July	31
February	28 (or 29 in a leap year)	August	31
March	31	September	30
April	30	October	31
May	31	November	30
June	30	December	31

his reformed calendar the years ending in two noughts (e.g. 1700, 1800, etc.) are only leap years if they are divisible by 400.

This system, called the *Gregorian calendar*, is the one in general use today. According to it 400 civil years contain $(400 \times 365) + 100 - 3 = 146\,097$ days, so that the average length of the civil year is $146\,097/400 = 365.2425$ days, a very good approximation indeed to the length of the tropical year.

2 The date of Easter

Easter day, the date to which such moveable feasts as Whitsun and Trinity Sunday are fixed, is usually the first Sunday after the fourteenth day after the first new Moon after March 21st. (For a more precise definition see *The Explanatory Supplement to the Astronomical Ephemeris and American Ephemeris and Nautical Almanac.*) You can find the date of Easter Sunday by the method and tables given, for example, in the *Book of Common Prayer*, 1662, or by one of several methods devised by various mathematicians over the centuries. Here I shall describe a method devised in 1876 which first appeared in *Butcher's Ecclesiastical Calendar*, and which is valid for all years in the Gregorian calendar, that is from 1583 and onwards. It makes repeated use of the result of dividing one number by another number, the integer part being treated separately from the remainder. A calculator displays the result of such a division as a string of numbers before and after a decimal point. The numbers appearing before the decimal point constitute the integer part; the numbers after the decimal point constitute the fractional part. The remainder may be found from the latter by multiplying it by the divisor (i.e. the number you have just divided by) and rounding the result to the nearest integer value. For example, 2000/19 = 105.263 157 9. The integer part is 105 and the fractional part is 0.263 157 9. Multiplying this by 19 gives 5.000 000 100 so that the remainder is 5.

I shall illustrate the method by calculating the date of Easter Sunday in the year 2000.

Method			*Example*
	Integer part	*Remainder*	
1. Divide the year by 19.	–	a	$\frac{2000}{19} = 105.263\,157\,9$ $a = 5$
2. Divide the year by 100.	b	c	$\frac{2000}{100} = 20.000\,000$ $b = 20$ $c = 0$
3. Divide b by 4.	d	e	$d = 5$ $e = 0$
4. Divide $(b+8)$ by 25.	f	–	$f = 1$
5. Divide $(b-f+1)$ by 3.	g	–	$g = 6$
6. Divide* $(19a+b-d-g+15)$ by 30.	–	h	$(19a+b-d-g+15) = 119$ $h = 29$
7. Divide c by 4.	i	k	$i = 0$ $k = 0$
8. Divide $(32+2e+2i-h-k)$ by 7.	–	l	$(32+2e+2i-h-k) = 3$ $l = 3$
9. Divide $(a+11h+22l)$ by 451.	m	–	$(a+11h+22l) = 390$ $m = 0$
10. Divide $(h+l-7m+114)$ by 31	n	p	$(h+l-7m+114) = 146$ $n = 4$ $p = 22$
11. Day of the month on which Easter Sunday falls is $p+1$. Month number is n (= 3 for March and = 4 for April).			$p+1 = 23$
∴ Easter Sunday 2000 is			**23rd April**

* $19a$ means 19 multiplied by a ($19 \times 5 = 95$ in this example).

3 Converting the date to the day number

In many astronomical calculations, we need to know the number of days in the year up to a particular date. We shall choose our starting point as 0 hours on January 0th, equivalent to the midnight between December 30th and 31st of the previous year; this may seem rather peculiar at first but as it simplifies the calculations we shall adopt it for our purposes. Midday on January 3rd is expressed as January 3.5 because three and a half days have elapsed since January 0.0. This is illustrated in Figure 1.

Finding the day number from the date is then a simple matter. Proceed as follows:

1. For every month up to, but not including, the month in question add the appropriate number of days according to Table 1. These totals are listed in Table 2*b*.
2. Add the day of the month.

For example. Calculate the day number of February 17th (not a leap year).

Day number = 31 + 17 = 48.

If you own a programmable calculator, you may be able to use the routine R1 to write a program enabling you to carry out the calculation automatically.

Figure 1. Defining the epoch.

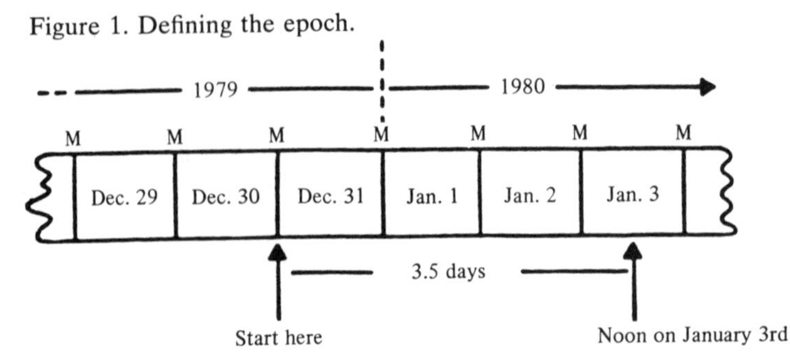

M = midnight

Table 2a. Days to the beginning of the year since epoch 1980 January 0.0

*1980:	0	1990:	3653
1981:	366	1991:	4018
1982:	731	*1992:	4383
1983:	1096	1993:	4749
*1984:	1461	1994:	5114
1985:	1827	1995:	5479
1986:	2192	*1996:	5845
1987:	2557	1997:	6210
*1988:	2922	1998:	6575
1989:	3288	1999:	6940

* Denotes a leap year.

Table 2b. Days to the beginning of the month

	Ordinary year	Leap year
January:	0	0
February:	31	31
March:	59	60
April:	90	91
May:	120	121
June:	151	152
July:	181	182
August:	212	213
September:	243	244
October:	273	274
November:	304	305
December:	334	335

Later on in this book we adopt the date 1980 January 0.0 as the starting point, or epoch, from which to calculate orbital positions. Days elapsed since this epoch at the beginning of each year up to 1999 are tabulated in Table 2*a*. To find the total number of days elapsed since the epoch simply add the appropriate number to the day number calculated in the previous paragraph.

For example. The number of days elapsed since the epoch on February 17th 1985 is 48 + 1827 = 1875.

This calculation may also be done via the Julian day number; see section 4.

Routine R1. Converting the date to the day number.

1. Key in the month number (e.g. 11 for November).
2. Is it greater than 2?
   - If yes, go to step 8.
   - If no, proceed with step 3.
3. Subtract 1 from month number.
4. Multiply by 63 (or 62 in a leap year).
5. Divide by 2.
6. Take the integer part.
7. Go to step 12.
8. Add 1 to month number.
9. Multiply by 30.6.
10. Take the integer part.
11. Subtract 63 (or 62 in a leap year).
12. Add the day of the month. The result is the day number.

* Sometimes the *modified Julian date*, MJD, is quoted. This is equal to JD − 2 400 000.5; MJD zero therefore began at 0h on November 17th 1858.

4 Julian day numbers

It is sometimes necessary to express an instant of observation as so many days and a fraction of a day after a given fundamental epoch. Astronomers have chosen this fundamental epoch as the Greenwich mean noon of January 1st 4713 B.C., that is midday as measured on the Greenwich meridian on January 1st of that year. The number of days which have elapsed since that time is referred to as the *Julian day number*, or *Julian date.** It is important to note that each new Julian day begins at 12h 00m GMT (UT), half a day out of step with the civil day.

The Julian date of any day in the Julian or Gregorian calendars may be found by the method given below. If you are working with years 'before Christ', count them logically, that is, the year before the year A.D. 1 is zero and the year preceding that is −1. As an example, we shall calculate the Julian date corresponding to 1985 February 17.25 (i.e. 6 a.m. on February 17th).

Method	*Example*
1. Set y = year, m = month and d = day.	$y = 1985$ $m = 2$ $d = 17.25$
2. If $m = 1$ or 2 subtract 1 from y and add 12 to m. Otherwise $y' = y$ and $m' = m$.	$y' = 1984$ $m' = 14$
3. If the date is later than 1582 October 15 (i.e. Gregorian calendar) calculate: (i) A = integer part of $(y'/100)$; (ii) $B = 2 - A +$ integer part of $(A/4)$. Otherwise $B = 0$.	$A = \text{INT}(1984/100)$ $= 19$ $B = 2 - 19 + \text{INT}(19/4)$ $= -13$
4. Calculate C = integer part of $(365.25 \times y')$.	$C = \text{INT}(365.25 \times 1984)$ $= 724\,656$
5. Calculate D = integer part of $(30.6001 \times (m' + 1))$.	$D = \text{INT}(30.6001 \times 15)$ $= 459$
6. Find $\text{JD} = B + C + D + d + 1\,720\,994.5$. This is the Julian date.	JD = **2 446 113.75**

The Julian date of the epoch 1980 January 0.0 is 2 444 238.5. We can easily find the number of days which have elapsed since the epoch by subtracting this number from the Julian date.

For example. The number of days elapsed since the epoch to 1985 February 17.0 is 2 446 113.5 − 2 444 238.5 = 1875, as found in the previous section.

5 Converting the Julian day number to the calendar date

It is sometimes necessary to convert a given Julian date into its counterpart in the Gregorian calendar. The method shown here works for all dates since January 1st 4713 B.C. For example, let us find the Gregorian date corresponding to JD = 2 446 113.75.

Method	*Example*
1. Add 0.5 to JD. Set I = integer part and F = fractional part.	JD = 2 446 113.75 +0.5 = 2 446 114.25 I = 2 446 114 F = 0.25
2. If I is larger than 2 299 160, calculate:	
(i) A = integer part of $\left(\frac{I - 1\,867\,216.25}{36\,524.25}\right)$;	A = 15.0
(ii) $B = I + 1 + A -$ integer part of $(A/4)$. Otherwise, set $A = I$.	B = 2 446 127.0
3. Calculate $C = B + 1524$.	C = 2 447 651.0
4. Calculate D = integer part of $\left(\frac{C - 122.1}{365.25}\right)$.	D = 6 700.0
5. Calculate E = integer part of $(365.25 \times D)$.	E = 2 447 175.0
6. Calculate G = integer part of $\left(\frac{C - E}{30.6001}\right)$.	G = 15.0
7. Calculate $d = C - E + F -$ integer part of $(30.6001 \times G)$. This is the day of the month (including the decimal fraction of the day).	d = **17.25**
8. Calculate $m = G - 1$ if G is less than 13.5, or $m = G - 13$ if G is more than 13.5. This is the month number.	m = **2**
9. Calculate $y = D - 4716$ if m is more than 2.5, or $y = D - 4715$ if m is less than 2.5. This is the year.	y = **1985**

Hence the Gregorian date is 1985 February 17.25.

6 Finding the day of the week

It is sometimes useful to know on what day of the week a particular date will fall. This can be found very easily once the Julian date has been calculated. Here, we shall find the day of the week corresponding to February 17th 1985.

Method	*Example*
1. Find the Julian day number corresponding to 0h UT on the day in question (§ 4).	1985 February 17.0 JD = 2 446 113.5
2. Calculate $A = \left(\frac{\text{JD}+1.5}{7}\right)$.	$A =$ 349 445.0000
3. Take the fractional part of A, multiply by 7, and round to the nearest integer. This is the weekday number, n. The day of the week is then as follows: Sunday, $n = 0$; Monday, $n = 1$; Tuesday, $n = 2$; Wednesday, $n = 3$; Thursday, $n = 4$; Friday, $n = 5$; Saturday, $n = 6$.	Fractional part = 0.0 $n =$ **0** **Sunday**

7 Converting hours, minutes and seconds to decimal hours

Most times are expressed as hours and minutes, or hours, minutes and seconds. For example, twenty to four in the afternoon may be written as 3.40 p.m., or 3h 40m p.m., or on a 24-hour clock as 15h 40m. In calculations, however, the time needs to be expressed in decimal hours on a 24-hour clock. The method of converting a time expressed in the format hours, minutes and seconds into decimal hours is given below. Some calculators have special keys to do this for you automatically.

Method	*Example*
	6h 31m 27s p.m.
1. Take the number of seconds and divide by 60.	27/60 = 0.450 00
2. Add this to the number of minutes and divide by 60.	31.45/60 = 0.524 17
3. Add the result to the number of the hours.	= 6.524 17
4. If the time has been given on a 12-hour clock, and it is p.m., add 12.	+12.0 = **18.524 17 hours**

8 Converting decimal hours to hours, minutes and seconds

When the result of a calculation is a time, it is normally expressed as decimal hours, and we need to convert it to hours, minutes and seconds. The method of doing so is given below. Again, some calculators have special keys to carry out this function automatically.

Method	*Example*
	18.524 17 hours
1. Take the fractional part and multiply by 60. The integer part of the result is the number of minutes.	0.524 17 × 60 = **31**.4502
2. Take the fractional part of the result and multiply by 60. This gives the number of seconds.	0.4502 × 60 = **27**.012
	18h 31m 27s

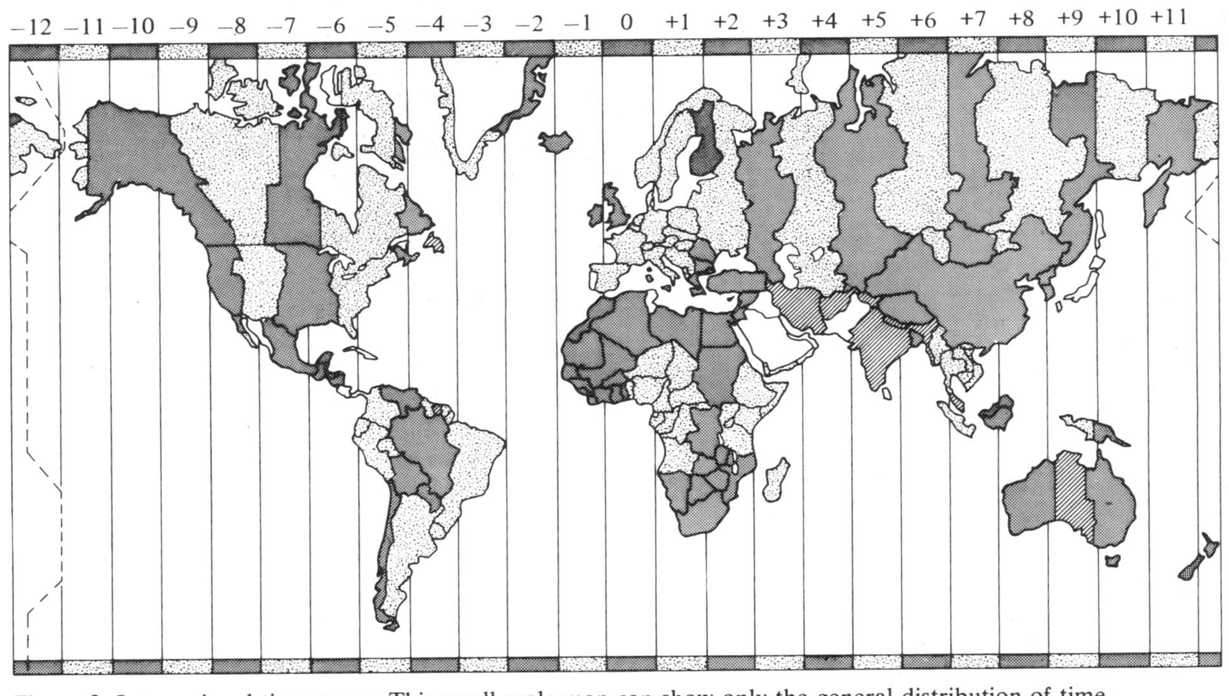

Figure 2. International time zones. This small-scale map can show only the general distribution of time zones around the world. If you are unsure of your own zone correction, you can check it by tuning your short-wave radio to the BBC World Service and comparing your watch with the GMT time pips broadcast every hour from London.

9 Converting the local time to GMT (UT)

Greenwich mean time (GMT), or universal time (UT) as it is also known, is related to the motion of the Sun as observed on the Greenwich meridian, longitude 0°. This time is used as the local civil time in Britain during the winter months but during the summer one hour is added to GMT to form British summer time (BST) so that the working day fits more conveniently into the daylight hours. Many other countries adopt a similar arrangement; sometimes the converted time is known as daylight saving time.

Countries lying on meridians east or west of the Greenwich meridian do not use GMT as their local civil time. It would obviously be impractical to do so as the local noon, the time at which the Sun reaches its maximum altitude, gets earlier or later with respect to the local noon on the Greenwich meridian as one moves east or west respectively. To avoid confusion, the world is divided into time zones, each zone usually corresponding to a whole number of hours before or after GMT, and small countries, or parts of large countries lying within a zone, adopt the zone time as their local civil time (see Figure 2). It is often convenient in making astronomical calculations to use GMT, and the local civil time may be converted into GMT in the following manner. For an example we convert daylight saving time 03h 37m on longitude 64° E (zone +4 hours) to GMT.

Method	*Example*
1. Convert local civil time to zone time if necessary to correct for daylight saving. Convert to decimal hours (§ 7).	03h 37m − 01h 00m = 02h 37m
	Zone time = 2.616 667 hours
2. Subtract zone correction.	− 4.00
3. If the answer is greater than 24, subtract 24. If the answer is negative, add 24. Convert back to hours, minutes and seconds (§ 8).	= −1.383 333 hours + 24.00 = 22.616 667 GMT = **22h 37m**

10 Converting GMT to local civil time

Given the time as UT or GMT, what is the corresponding local civil time? For example, what is the local civil time on longitude 64° E (zone +4 hours) during daylight saving when the GMT is 22h 37m?

Method	*Example*
1. Convert GMT to decimal hours (§ 7).	22h 37m = 22.616 667 hours
2. Add the zone correction. If the answer is greater than 24, subtract 24. If the answer is negative, add 24.	+ 4.00 = 26.616 667 hours – 24.00 = 2.616 667
3. Convert to hours, minutes and seconds (§ 8) and correct for daylight saving (if necessary). This is the local civil time.	= 02h 37m + 01h 00m = **03h 37m**

11 Sidereal time (ST)

Greenwich mean time, and therefore the local civil time in any part of the world, is regulated by the motion of the Sun. Roughly speaking, we may take one solar day as the time between two successive passages of the Sun across the meridian as observed at a particular place. Astronomers are interested, however, in the motion of the stars; in particular they need to use a clock whose rate is such that any star is observed to return to the same position in the sky after exactly 24 hours have elapsed according to the clock. Such a clock is called a sidereal clock and its time, being regulated by the stars, is called sidereal time (ST). Solar time, of which GMT is an example, is not the same as sidereal time because during the course of one solar day the Earth moves nearly one degree along its orbit round the Sun. Hence, the Sun appears progressively displaced against the background of stars when viewed from the Earth; turning that around, the stars appear to move with respect to the Sun. Any clock, therefore, which keeps time by the Sun does not do so by the stars.

There are about $365\frac{1}{4}$ solar days in the year,* the time taken by the Sun to return to the same position with respect to the background of stars. During this period, the Earth makes about $366\frac{1}{4}$ revolutions about its own axis; there are therefore this many sidereal days in the year. Each sidereal day is thus slightly shorter than the solar day, 24 hours of sidereal time corresponding to 23h 56m of solar time. Greenwich mean time and Greenwich sidereal time agree at one instant every year at the autumnal equinox (around September 22nd). Thereafter, the difference between them grows in the sense that ST runs faster than GMT, until exactly half a year later it is 12 hours. After one year, the times again agree.

The formal definition of sidereal time is that it is the hour-angle of the vernal equinox (see section 18).

* See the definition of year in the Glossary.

12 Conversion of GMT to GST

This section describes a simple procedure by which the Greenwich mean time may be converted to Greenwich mean sidereal time (GST). It is accurate to better than one tenth of a second.

The constants A and C are independent of the year and are listed in Table 3. The constant B does, however, depend on the year. The values for the years 1975 to 2000 are given in Table 3. To find B accurately for any other year, you will need to look up the Greenwich mean sidereal time at 0h UT on January 0th of the year in question in the *Astronomical Ephemeris*. Convert this time to decimal hours (section 7) and subtract it from 24. The result is constant B. Alternatively, you can calculate B by the method given at the end of this section.

Table 3

A: 0.065 709 8 C: 1.002 738 D: 0.997 270

B:

1975:	17.397 610	1990:	17.373 487
1976:	17.413 525	1991:	17.389 402
1977:	17.363 730	1992:	17.405 316
1978:	17.379 643	1993:	17.355 521
1979:	17.395 558	1994:	17.371 435
1980:	17.411 472	1995:	17.387 349
1981:	17.361 677	1996:	17.403 263
1982:	17.377 592	1997:	17.353 468
1983:	17.393 506	1998:	17.369 382
1984:	17.409 421	1999:	17.385 297
1985:	17.359 625	2000:	17.401 211
1986:	17.375 539		
1987:	17.391 453		
1988:	17.407 368		
1989:	17.357 573		

Example. What was the GST at 14h 36m 51.67s on April 22nd 1980?

Method	*Example*
1. Find the number of days between January 0.0 and the date in question (§ 3).	Days = 113.0
2. Multiply by constant A.	× 0.065 709 8
	= 7.425 207
3. Subtract constant B. The result is T_0.	− 17.411 472
	$T_0 =$ −9.986 265
4. Convert GMT to decimal hours (§ 7).	GMT = 14.614 353
5. Multiply by constant C.	× 1.002 738
6. Add this to T_0. If the result is more than 24, subtract 24. If the result is negative, add 24. This is GST in hours.	= 14.654 367
	+ −9.986 265
	GST = 4.668 102
7. Convert result to hours, minutes and seconds (§ 8).	GST = **04h 40m 5.17s**

To calculate constant B:

Method	*Example*
	e.g. for year 1979
1. Calculate the Julian date on January 0.0 of year in question (§ 4).	JD = 2 443 873.50
2. Calculate $S = \text{JD} - 2\,415\,020.0$.	$S =$ 28 853.50
3. Calculate $T = S/36\,525.0$.	$T = 0.789\,965\,777$
4. Calculate $R = 6.646\,065\,6 + (2400.051\,262 \times T) + (0.000\,025\,81 \times T^2)$.	$R = 1\,902.604\,442$
5. Calculate $U = R - (24 \times (\text{year} - 1900))$.	$U = 6.604\,442$
6. Subtract U from 24. This is B.	$B =$ **17.395 558**

13 Conversion of GST to GMT

Here we deal with the reverse problem of the previous section, namely that of converting a given Greenwich mean sidereal time into the corresponding Greenwich mean time (UT). The problem is complicated, however, by the fact that the sidereal day is slightly shorter than the solar day so that on any given date a small range of sidereal times occurs twice. This range is about four minutes long, the sidereal times corresponding to GMT 0h to 0h 04m occurring again from GMT 23h 56m to midnight (see Figure 3). The method given here correctly converts sidereal times in the former interval, but not in the latter.

Constants *A* and *B* are as in section 12. Constant *D* is independent of the year and is listed in Table 3. The accuracy of this method is the same as that of section 12, namely better than one tenth of a second.

Figure 3. GMT and GST for April 22nd 1980. The shaded intervals of GST occur twice on the same day.

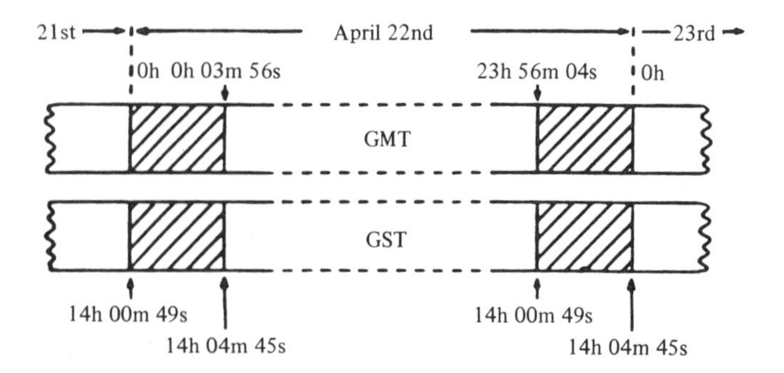

Example. At GST = 04h 40m 5.17s on April 22nd 1980, what was the GMT?

Method	*Example*
1. Find the number of days between January 0.0 and the date in question (§ 3).	Days = 113.0
2. Multiply by constant A.	× 0.065 709 8
	= 7.425 207
3. Subtract constant B. If the result is negative, add 24. This is T_0.	− 17.411 472
	= −9.986 265
	+ 24.0
	T_0 = 14.013 735
4. Convert GST to decimal hours (§ 7).	GST = 4.668 103
5. Subtract T_0. If the result is negative, add 24.	− 14.013 735
	= −9.345 633
	+ 24.0
	= 14.654 367
6. Multiply by constant D. This is the GMT in hours.	× 0.997 270
	= 14.614 361
7. Convert the result to hours, minutes and seconds (§ 8).	GMT = **14h 36m 51.70s**

14 Local sidereal time (LST)

The Greenwich sidereal time discussed in the previous sections is the sidereal time correct for observations made on the Greenwich meridian, longitude 0°. It is the local sidereal time for the Greenwich meridian. As you move west or east from longitude 0°, however, the local sidereal time gets earlier or later respectively because the hour-angle of the vernal equinox, which defines the local sidereal time, changes. You can calculate your local sidereal time, given the Greenwich sidereal time, very easily as the difference between the two times in hours is simply the geographical longitude in degrees divided by 15. Longitudes west give local sidereal times earlier than GST and longitudes east later. Take the example: what is the local sidereal time on the longitude 64° W when the Greenwich sidereal time is 4h 40m 5.17s.

Method	*Example*
1. Convert GST to decimal hours (§ 7).	GST = 4.668 103 hours
2. Convert longitude difference in degrees to difference in time by dividing by 15.	64° = 4.266 667 hours
3. If the longitude is W, subtract. If the longitude is E, add. If the result is greater than 24, subtract 24. If the result is negative, add 24. This is the LST in hours.	LST = 0.401 436 hours
4. Convert LST to hours, minutes and seconds (§ 8).	LST = **0h 24m 5.17s**

15 Converting LST to GST

This problem is the reverse of that treated in section 14, namely given the local sidereal time at a particular place what is the corresponding Greenwich sidereal time. As an example, we shall calculate the GST when the LST on longitude 64° W is 0h 24m 5.17s.

Method	*Example*
1. Convert the LST to decimal hours (§ 7).	LST = 0.401 436 hours
2. Convert the longitude difference in degrees to difference in time by dividing by 15.	64° = 4.266 667 hours
3. If the longitude is W, add. If the longitude is E, subtract. If the result is greater than 24, subtract 24. If the result is negative, add 24. This is the GST in hours.	GST = 4.668 103 hours
4. Convert GST to hours, minutes and seconds (§ 8).	GST = **04h 40m 5.17s**

16 Ephemeris time (ET)

Universal time and sidereal time are both tied directly to the period of rotation of the Earth about its polar axis. The Earth is being used in effect as the balance-wheel of a cosmic clock whose tick defines the length of the day. With the advent of extremely accurate atomic clocks, however, it has become apparent that the Earth's rotation is not strictly uniform but shows small erratic fluctuations which are not well understood. UT and ST, being reckoned by this irregular cosmic clock, are therefore not strictly uniform either. Astronomers, however, need a system of time measurement which *is* uniform since the theories of celestial mechanics assume that such a quantity exists. For example, two solid bodies in orbit about one another far away from any external influence should have an unchanging orbital period when measured on a regular clock. Astronomers therefore use *ephemeris time* (ET), which

Figure 4. The variation of $\Delta T = \text{ET} - \text{UT}$ since 1800.

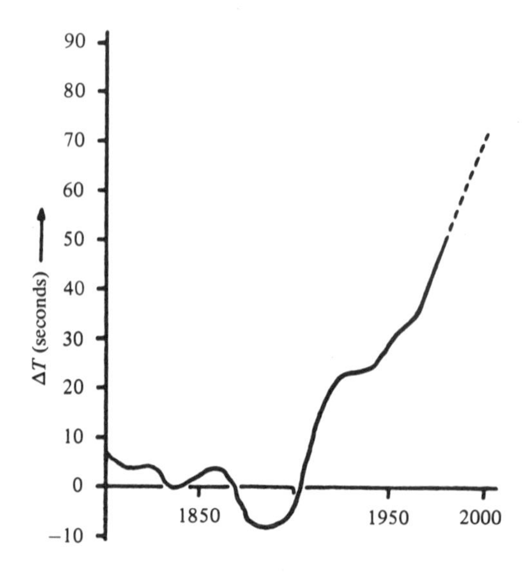

is calculated from the motion of the Moon and assumed to be uniform. It has been chosen to agree as nearly as possible with the measure of universal time (GMT) during the nineteenth century, and it is unlikely that the two measures will differ by more than a few minutes in the twentieth.

The primary unit of ephemeris time is the length of the tropical year at 1900 January 0.5 ET which contains 31 556 925.974 7 ephemeris seconds. We need not be too concerned by all this since very high accuracy is not the aim of this book. In almost every case we can take ET = UT = GMT without noticing the difference. Only when calculating the motion of the Moon, and predicting eclipses, will it pay us to take account of the difference between ET and UT. In January 1980 this difference was about 51 seconds, UT being behind ET; that is, ET − UT = ΔT = 51 seconds (January 1980).

Figure 4 shows how ΔT has varied since 1800; we can predict that its value in the year 2000 might be +70s, but only direct observations at that time will confirm this.

Coordinate systems

To fix the position of any astronomical object, we must have a frame of reference, or coordinate system, which assigns a different pair of numbers to every point in the sky. The two numbers, or coordinates, usually refer to 'how far round' and 'how far up', just as do the longitude and latitude of an object on the Earth's surface. There are several such coordinate systems which you may meet, and we shall be concerned with four of these, namely the horizon system, the equatorial system, the ecliptic system and the galactic system. Each system takes its name from the fundamental plane which it uses as a reference; for instance, the ecliptic coordinate system makes all its measurements with respect to the plane of the ecliptic, the plane of the Earth's orbit about the Sun. In the next few sections, we shall find how to convert any position given in one system into the equivalent coordinates of another system. We shall also find how to describe positions on the surface of the Sun, how to deal with the problems of calculating the times of rising and setting, and with the effects of the Earth's precession, atmospheric refraction, and parallax on the apparent position of a celestial body.

17 Horizon coordinates

The horizon coordinates, azimuth and altitude, of an object in the sky are referred to the plane of the observer's horizon (see Figure 5). Imagine an observer standing at point O; then his horizon is the circle NESW, where the letters refer to the north, east, south and west points of his horizon respectively. The direction north, by the way, relates to the direction of the north pole on the Earth's rotation axis and not to the magnetic north pole. You must imagine the stars as fixed on the surface of the hemisphere with the observer at the centre as in Figure 5; the whole sphere of which this hemisphere is part is called the *celestial sphere*. The point Z directly over the observer's head is called the *zenith*; the direction OZ is the direction defined by a plumb line held by the observer. Now consider a star X and imagine a great circle (i.e. a circle drawn on the surface of the sphere whose centre is the same as that of the sphere) going through Z and X; it meets the horizon at point B. The *altitude*, a, of the star is then the angle subtended at O by the points X and B. The *azimuth*, A, is the angle subtended by the points N and B. Hence, the altitude is 'how far up' in degrees (negative if below the horizon) and the azimuth is 'how far round' from the north direction, also measured in degrees. A increases from 0° to 360° as you go around in the sense NESW, N being 0°, S being 180°, etc.

Figure 5. Horizon coordinates.

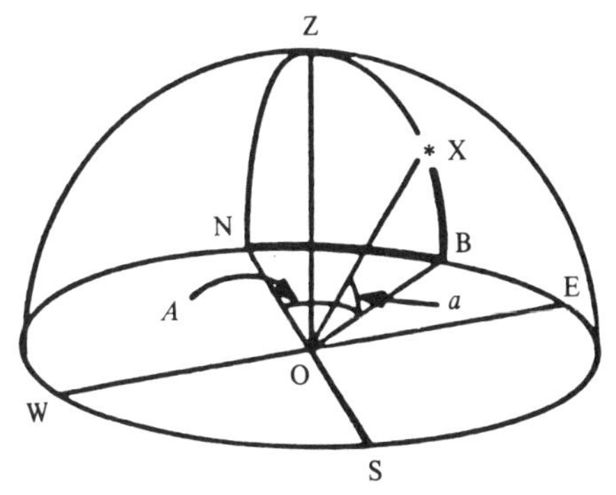

The altitudes and azimuths of all heavenly bodies except geostationary satellites are continually changing with time as the Earth rotates. This coordinate system then, marvellous for setting the direction of your telescope, is not much good for fixing the positions of the stars. Another frame of reference is needed to do that, which is independent of the Earth's motion. It is described in the next section.

18 Equatorial coordinates

As their name suggests, these coordinates are referred to the plane of the Earth's equator (see Figure 6*a*). The observer is at O and the plane containing the circle NESW is again his horizon with Z his zenith point. You are to imagine now that the figure represents the view obtained at a vast distance from the Earth. The planet, with the observer standing on it, has shrunk to a tiny dot at the centre of the diagram, but the plane of the equator has been extended to cut the celestial sphere along the circle E♈RW. This is the equatorial plane and is inclined at the angle $90° - \phi$ to the horizon where ϕ is the observer's geographical latitude. For observations at latitude 52° N this angle is 38°. At right angles to the equatorial plane along the line OP lies the axis of rotation of the Earth; it intersects the celestial sphere at P, the *north celestial pole*, or *north pole* for short. Since this is the line about which the Earth spins, all the stars appear to describe circles in the sky about P.

Figure 6*b* shows the situation as seen by the observer O looking up into the sky. The south point, S, of his horizon is marked and so is the imaginary trace of the equator, C♈RD. The arc extending down through R and S is the great circle which goes through NPZRS in Figure 6*a*. The arc extending down through XC is another great circle, not marked in Figure 6*a*, which goes through PXC. Consider the star at X. The arc XC, or the angle subtended at O by the points X and C, is called the *declination*, δ, of X, defining 'how far up' from, or north of, the equator. The other coordinate, 'how far round', is defined with respect to a fixed direction in the sky, marked by the symbol ♈. This direction, called the *vernal equinox* or the *first point of Aries*, lies along the line of the intersection of the plane of the Earth's equator with that of the Earth's orbit around the Sun. But we needn't worry about such definitions at the moment. All we need to know is that the direction ♈ remains fixed with respect to the stars (but see section 33), and that we measure the other coordinate with respect to it. This

Figure 6. Equatorial coordinates: (*a*) on the celestial sphere, (*b*) as seen from the ground.

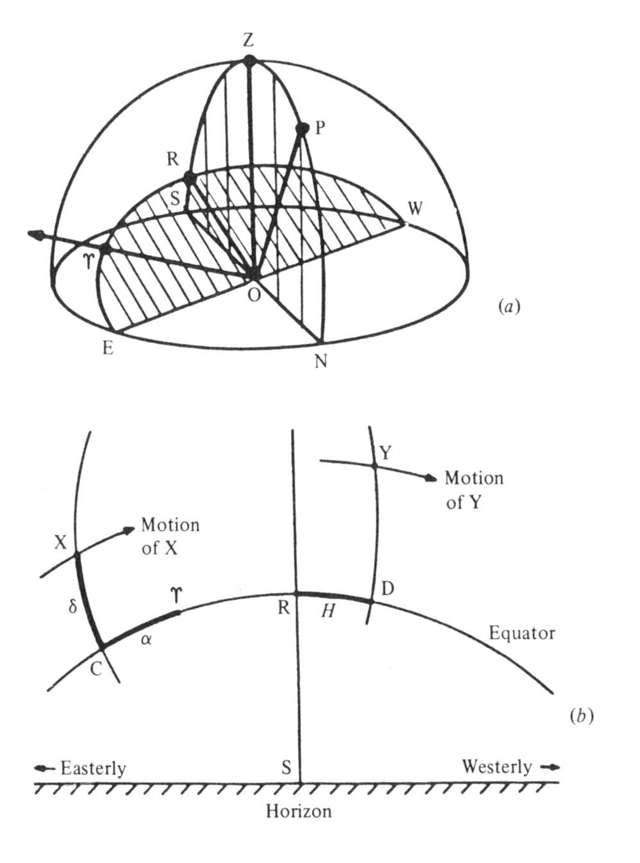

coordinate is called the *right ascension*, α, and is the angle subtended at O by the points ♈ and C.

Throughout the course of the day the star X moves steadily westwards along a circle centred on P, completing one revolution in 24 hours of sidereal time (see section 11). Since this circle is a parallel circle to that of the equator the declination does not change. Furthermore, since the direction ♈ is fixed in the heavens, it appears to move along the equator at exactly the same rate as X moves along the circle. Hence the right ascension does not change either. Thus α and δ are ideal coordinates for describing the positions of the stars and other 'fixed' heavenly bodies.

Related to the right ascension is another 'how far round' coordinate called the *hour-angle*, H (see Figure 6*b*). For the star Y it is defined as the angle subtended at O by the points R and D and is a measure of how far the star has travelled along the equator from the southern point R, that is, a measure of the time since it crossed the meridian. H obviously increases uniformly as the day proceeds; when H is zero, the star crosses the great circle NPZRS (Figure 6*a*). This circle is called the *meridian* and the star is said to *transit* or *culminate*. Its altitude (section 17) is then maximum and its azimuth* is 180° (provided that its declination is less than the geographical latitude).

The declination is measured in degrees, positive north of the equator and negative south of it. The hour-angle and the right ascension may also be measured in degrees, 0° to 360°. α is measured in the sense that it increases as you move *east* from ♈; the point ♈ itself is at 0°. (Note that this is in the *opposite* sense to that in which H is measured.) More commonly, however, these two coordinates H and α are measured in hours, minutes and seconds of time from 0 to 24 hours. One complete revolution, 360°, corresponds to 24 hours of sidereal

* Some authors measure azimuth from the south point rather than the north point, in which case $A = 0°$ at transit.

time; thus one hour is equivalent to 15°. The two statements 'the right ascension of X is 90°' and 'the right ascension of X is 6h' are entirely equivalent. To convert from one to the other simply multiply or divide by 15.

A useful result of measuring the right ascension in time is that the star transits when the local sidereal time is equal to the right ascension.

Finally, a word more about hour-angles. With a little thought you can quickly see that if a star's azimuth is west, that is if A is greater than 180°,* the hour-angle is between 0h and 12h (and vice versa); and if the star's azimuth is east, that is if A is less than 180°,* the hour-angle is between 12h and 24h.

19 Ecliptic coordinates

The plane containing the Earth's orbit around the Sun is called the *ecliptic* and the other planets in our Solar System also move in orbits close to this plane. When making calculations on objects in the Solar System it is therefore often convenient to define positions with respect to the ecliptic, that is, to use the ecliptic coordinate system. This system, like the equatorial system described in section 18 above, also uses the vernal equinox, ♈, as its reference direction. Figure 7, which is similar to Figure 6*b*, shows how it goes.

Figure 7. Ecliptic coordinates as seen from the ground.

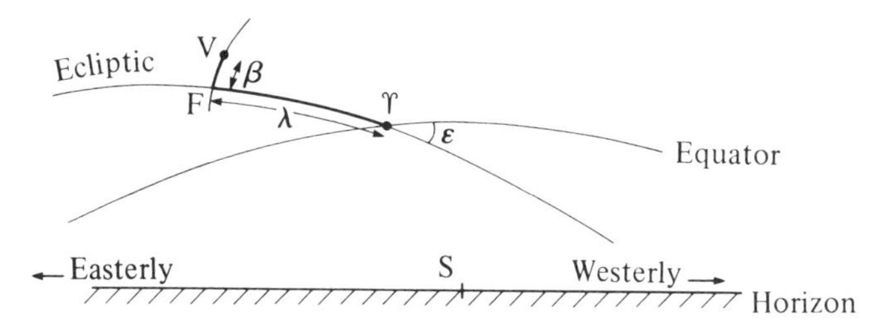

* This applies for stars whose declinations are less than the geographical latitude.

The imaginary traces of the planes of the equator and the ecliptic are drawn on the sky, and their point of intersection is the vernal equinox, ♈. The two planes are inclined to each other at an angle of about 23° 26′, called the *obliquity* of the ecliptic and given the symbol ε. This angle is the Earth's tilt from the perpendicular to the plane of the ecliptic. Also marked in Figure 7 is a planet, V. Part of the trace of the imaginary great circle from the pole of the ecliptic (i.e. the point where the line drawn through the Sun perpendicular to the ecliptic meets the celestial sphere) down through V is marked and this cuts the ecliptic at F. Then the *ecliptic longitude*, λ, of V is defined to be the angle subtended by the points F and ♈, and the *ecliptic latitude*, β, the angle subtended by the points F and V.

As with equatorial coordinates, β is positive if the planet is above (i.e. north of) the ecliptic and negative below it. The sense of λ is such that λ increases as you move eastwards along the ecliptic. Both λ and β are usually measured in degrees.

During the course of the year the Sun moves eastwards along the trace of the ecliptic. By definition, its ecliptic latitude is always zero. On March 21st, it is at the position ♈ and its right ascension and declination are both zero. Its ecliptic longitude is also zero. Thereafter, its ecliptic longitude steadily increases until three months later it is 90°, midsummer in the northern hemisphere. After the course of one year, the Sun has returned to its starting position having traversed 360° of ecliptic longitude.

20 Galactic coordinates

Astronomers occasionally need to describe the relations between stars or other celestial objects within our own Galaxy and to do so it is convenient to use the galactic coordinate system. This time, the fundamental plane is the plane of the Galaxy and the fundamental direction is the line joining our Sun to the centre of the Galaxy. Figure 8 describes the situation. The point marked S represents the Sun, G is the centre of the Galaxy, and X a star which does not lie in the galactic plane. In equatorial coordinates, the position of G is $\alpha = 17$h 42m and $\delta = -28.9$ degrees. The lines SG and SX′ both lie in the plane of the Galaxy; the point X′ is the projection of the star's position onto the plane. The *galactic longitude* is defined to be the angle l measured in the plane, and the *galactic latitude* is defined to be the angle b measured perpendicular to it. The longitude increases from 0° to 360° in the same direction as increasing right ascension, while the latitude ranges from 0° to 90° north of the plane and from 0° to −90° south of it. These coordinates may be used, for example, to find the position of a star in the Milky Way.

Figure 8. Galactic coordinates.

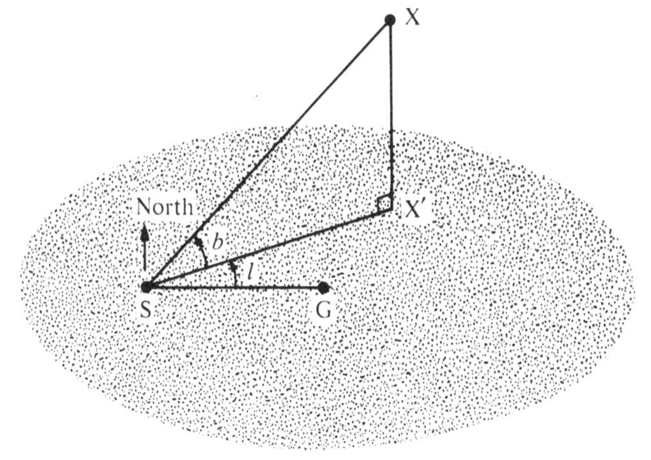

21 Converting between decimal degrees and degrees, minutes and seconds

Angles measured in degrees are usually expressed as degrees, minutes and seconds; the minutes and seconds are called minutes and seconds of arc to distinguish them from time. Calculations are best done, however, with decimal degrees and the methods of conversion between these two forms are exactly the same as the methods for conversion between hours, minutes and seconds and decimal hours. Thus the procedures listed in sections 7 and 8 may be used reading 'degrees' for 'hours'. Instruction 4 in section 7 obviously does not apply.

As an example, the angle $182° 31' 27''$ is equal to 182.5242 degrees.

22 Converting between angles expressed in degrees and angles expressed in hours

It is common astronomical practice to express the hour-angle or right ascension of a star in hours, minutes and seconds of time rather than in degrees. We can transform one to the other by noting that 360° of Earth's rotation takes place in one day, or 24 hours. Thus 360° is equivalent to 24 hours or 15° to one hour. Table 4 illustrates this equivalence more completely. To convert between angles expressed in *decimal* hours and angles expressed in *decimal* degrees, simply multiply or divide by 15. For example, the right ascension $\alpha = 09\text{h}\ 36\text{m}\ 10.2\text{s}$ is equivalent to $\alpha = 144°\ 02'\ 33''$.

Table 4. Expressing angles in degrees or time

Unit of time	Equivalent angle
1 day	360 degrees
1 hour	15 degrees
1 minute	15 arcmin
1 second	15 arcsec
Unit of angle	**Equivalent time**
1 radian	3.819 719 hours
1 degree	4 minutes
1 arcmin	4 seconds
1 arcsec	0.066 667 seconds

23 Converting between one coordinate system and another

It is very often necessary to convert the coordinates of a heavenly body expressed in one coordinate system into the equivalent coordinates of another system. This is the case when, for instance, you have found the position of a planet in ecliptic coordinates and you then wish to convert to horizon coordinates to see where to look in the sky. The formulae for conversion between the equatorial system and any of the other three systems, ecliptic, horizon or galactic, are relatively straightforward. The conversion, therefore, is usually best done via the equatorial system, as illustrated in Figure 9. The arrows indicate the conversions treated explicitly in this book in the section specified by the number at the side. For example, to convert from galactic coordinates to horizon coordinates first convert to equatorial coordinates (section 30) and then to horizon coordinates (section 25).

Figure 9. Converting between coordinate systems.

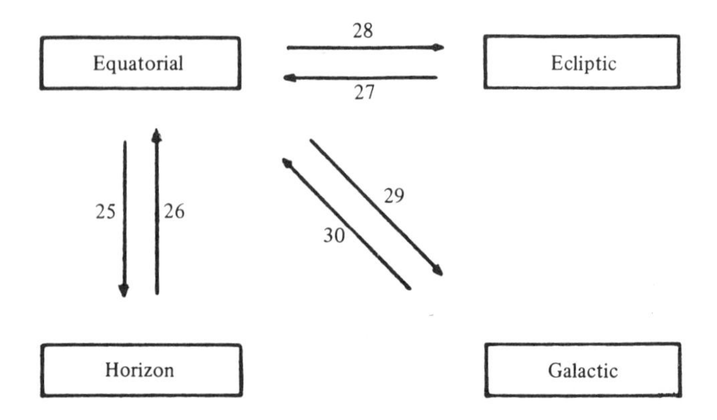

24 Converting between right ascension and hour-angle

The hour-angle, H, and the right ascension, α, are related by the simple formula

$$H = \text{LST} - \alpha,$$

where LST is the local sidereal time. Let us take as an example the problem of finding the local hour-angle of a star whose right ascension is α = 18h 32m 21s, at a point whose longitude is 64° W, on April 22nd 1980 at 14h 36m 51.67s GMT.

Method	*Example*
1. Convert GMT to GST (§ 12).	GST = 4.668 104 hours
2. Convert GST to LST (§ 14).	LST = 0.401 436 hours
3. Convert α into decimal hours (§ 7).	α = 18.539 167 hours
4. Subtract α from LST. If the result is negative, add 24. This is the hour-angle in decimal hours.	H = 5.862 269 hours
5. Convert H to hours, minutes and seconds (§ 8).	H = **05h 51m 44s**

Converting an hour-angle back to its equivalent right ascension is a very similar process. For example, what was the right ascension of the star whose hour-angle was 05h 51m 44s on April 22nd 1980 at 14h 36m 51.67s GMT, observed from longitude 64° W?

Method	*Example*
1. Convert GMT to GST (§ 12).	GST = 4.668 104 hours
2. Convert GST to LST (§ 14).	LST = 0.401 436 hours
3. Convert H to decimal hours (§ 7).	H = 5.862 269 hours
4. Subtract H from LST. If the result is negative, add 24. This is the right ascension in hours.	α = 18.539 167 hours
5. Convert α to hours, minutes and seconds (§ 8).	α = **18h 32m 21s**

25 Equatorial to horizon coordinate conversion

The formulae describing the relationships between hour-angle, H, declination, δ, azimuth, A, and altitude, a, are:

$$\sin a = \sin \delta \sin \phi + \cos \delta \cos \phi \cos H,$$

$$\cos A = \frac{\sin \delta - \sin \phi \sin a}{\cos \phi \cos a},$$

where ϕ is the observer's geographical latitude. (The hour-angle may be found from the right ascension by the method of section 24.) These may be dealt with in the following way using the example: what are the altitude and azimuth of a star whose hour-angle is 05h 51m 44s and declination is 23° 13′ 10″? The observer's latitude is 52° N.

Method	*Example*
1. Convert hour-angle to decimal hours (§ 7).	$H =$ 5.862 269 hours
2. Multiply by 15 to convert H to degrees (§ 22).	$H =$ 87.934 035 degrees
3. Convert δ into decimal degrees (§ 21).	$\delta =$ 23.219 444 degrees
4. Find $\sin a = \sin \delta \sin \phi + \cos \delta \cos \phi \cos H$.	$\sin a =$ 0.331 073
5. Take inverse sin to find a.	$a =$ 19.333 925 degrees
6. Find $\cos A = \dfrac{\sin \delta - \sin \phi \sin a}{\cos \phi \cos a}$.	$\cos A =$ 0.229 567
7. Take inverse cos to find A'.	$A' =$ 76.728 442
8. Find $\sin H$.	$\sin H =$ 0.999 350
If negative, the true azimuth is $A = A'$.	(positive)
If positive, the true azimuth is $A = 360 - A'$.	$A = 283.271\,558$ degrees
9. Convert a and A to degrees, minutes and seconds (§ 21).	$a =$ **19° 20′ 02″** $A =$ **283° 16′ 18″**

Step 8 is necessary because calculators can only return inverse trigonometrical functions correctly (inverse sin, inverse cos, inverse tan) over half the range of 0° to 360°. For example, try cos 147°. The answer is −0.8387 which reverts to 147° when you take inverse cos. But now try cos 213°. The answer is again −0.8387 which, when you take inverse cos, gives 147°. Hence, whenever the inverse is taken an ambiguity arises which has to be cleared up by another means.

Note that negative angles can be transformed back into the range 0° to 360° by simply adding 360. An example is −87.23 degrees which is the same as 360 − 87.23 = 272.77 degrees.

26 Horizon to equatorial coordinate conversion

This problem is the reverse of that of the preceding section, namely given a star's altitude, a, and azimuth, A, what are its declination, δ, and hour-angle, H? The appropriate formulae are:

$$\sin \delta = \sin a \sin \phi + \cos a \cos \phi \cos A,$$

$$\cos H = \frac{\sin a - \sin \phi \sin \delta}{\cos \phi \cos \delta},$$

where ϕ is the observer's geographical latitude. Notice that these formulae are exactly the same as those given in section 25 except that δ and H have been substituted for a and A and vice versa. This fact is useful when writing a program for a programmable calculator since exactly the same program can be used to convert δ, H to a, A or a, A to δ, H.

Let us take the example: a star is observed by an observer at latitude 52° N to have an altitude of 19° 20′ 02″ and an azimuth of 283° 16′ 18″. What are its hour-angle and declination? If the observer is on the Greenwich meridian and the GST is 0h 24m 05s what is the right ascension?

Method	*Example*
1. Convert azimuth to decimal degrees (§ 21).	$A = 283.271\,667$ degrees
2. Convert altitude to decimal degrees (§ 21).	$a = 19.333\,889$ degrees
3. Find $\sin \delta = \sin a \sin \phi + \cos a \cos \phi \cos A$.	$\sin \delta = 0.394\,255$
4. Take inverse sin to find δ.	$\delta = 23.219\,492$
5. Find $\cos H = \dfrac{\sin a - \sin \phi \sin \delta}{\cos \phi \cos \delta}$.	$\cos H = 0.036\,048$
6. Take inverse cos to find H'.	$H' = 87.934\,155$ degrees
7. Find sin A. If negative, the true hour-angle is $H = H'$. If positive, the true hour-angle is $H = 360 - H'$.	$\sin A = -0.973\,293$ (negative) $H = 87.934\,155$ degrees

Method (*continued*)	*Example*
8. Convert H into hours by dividing by 15 (§ 22).	$H =$ 5.862 277 hours
9. Convert H and δ into minutes and seconds form (§§ 8 and 21).	$H =$ **05h 51m 44s** $\delta =$ **23° 13′ 10″**

Again, step 7 is necessary to remove the ambiguity introduced by taking the inverse of cos.

The second part of the question, converting the hour-angle into the right ascension, may be done using the formula

$$\alpha = \mathrm{LST} - H,$$

where LST is the local sidereal time (=GST in this example), as shown in section 24. Thus:

Method	*Example*
1. Convert H into decimal hours (§ 7).	$H =$ 5.862 277 hours
2. Convert LST into decimal hours (§ 7).	LST = 0.401 389 hours
3. Subtract H from LST. If the result is negative, add 24. This is α in hours.	$\alpha =$ 18.539 112 hours
4. Convert α into hours, minutes and seconds (§ 8).	$\alpha =$ **18h 32m 21s**

Looking at the star atlas, we find a sixth-magnitude star in the constellation of Hercules listed near this position.

27 Ecliptic to equatorial coordinate conversion

The ecliptic longitude, λ, and the ecliptic latitude, β, may be converted into right ascension, α, and declination, δ, using the formulae:

$$\alpha = \tan^{-1}\left\{\frac{\sin\lambda\cos\varepsilon - \tan\beta\sin\varepsilon}{\cos\lambda}\right\},$$

$$\delta = \sin^{-1}\{\sin\beta\cos\varepsilon + \cos\beta\sin\varepsilon\sin\lambda\},$$

where ε is the obliquity of the ecliptic, the angle between the planes of the equator and the ecliptic. This angle changes slowly with time and for high accuracy the appropriate value should be used. If, for example, α and δ are referred to the standard epoch of 1950.0 (see section 33), then ε should have its 1950.0 value. In the examples given here and in the following section, we assume the 1980.0 value of $\varepsilon =$ 23.441 884 degrees. The method of calculating ε for any other epoch is given at the end of this section.

Our example this time is: what were the right ascension and the declination of a planet whose ecliptic coordinates were longitude 139° 41′ 10″ and latitude 4° 52′ 31″?

Method	*Example*
1. Convert λ and β into decimal degrees (§ 21).	$\lambda =$ 139.686 111 degrees $\beta =$ 4.875 278 degrees
2. Find $\sin\delta = \sin\beta\cos\varepsilon + \cos\beta\sin\varepsilon\sin\lambda$ (with $\varepsilon = 23.441\,884$ degrees).	$\sin\delta =$ 0.334 420
3. Take inverse sin to find δ in decimal degrees.	$\delta =$ 19.537 269 degrees
4. Find $y = \sin\lambda\cos\varepsilon - \tan\beta\sin\varepsilon$.	$y =$ 0.559 644
5. Find $x = \cos\lambda$.	$x =$ −0.762 512 $\dfrac{y}{x} =$ −0.733 948
6. Find $\alpha' = \tan^{-1}\left(\dfrac{y}{x}\right)$.	$\alpha' =$ −36.276 732 degrees
7. We have to remove the ambiguity which arises from taking the inverse tan. The rule is that α should lie in the quadrant indicated by the signs of x and y in Figure 10. Add or subtract 180 or 360 to α' to bring it into the correct quadrant, unless it is already there, in which case $\alpha = \alpha'$.	x negative y positive $\therefore \alpha = \alpha' + 180.0$ = 143.723 268 degrees
8. Convert α to hours by dividing by 15 (§ 22).	$\alpha =$ 9.581 551 hours
9. Convert α and δ to minutes and seconds form (§§ 21 and 8).	$\alpha =$ **09h 34m 53.6s** $\delta =$ **19° 32′ 14.2″**

Figure 10. Removing the ambiguity of taking $\tan^{-1} y/x$.

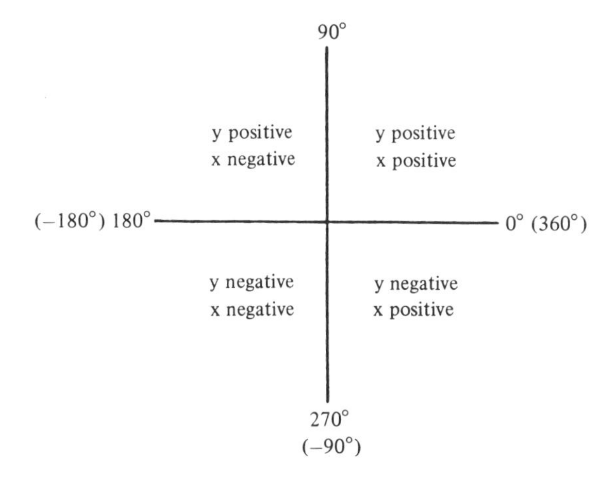

The value of the mean obliquity of the ecliptic is given by

$$\varepsilon = 23° 27' 08.26'' - 46.845''T - 0.0059''T^2 + 0.001\,81''T^3$$

where T is the number of Julian centuries since epoch 1900 January 0.5. For example, what was the mean obliquity of the ecliptic in 1980.0?

Method	*Example*
1. Calculate the Julian date (§ 4).	1980 January 0.0 JD = 2 444 238.50
2. Subtract 2 415 020.0 (=JD for 1900 January 0.5).	− 2 415 020.0 = 29 218.50 days
3. Divide by 36 525.0. The result is T.	T = 0.799 959 centuries
4. Calculate $\Delta\varepsilon = 46.845T + 0.0059T^2 - 0.001\,81T^3$.	$\Delta\varepsilon$ = 37.476 925 arcsec
5. Divide by 3600 to convert to degrees.	$\Delta\varepsilon$ = 0.010 410 degrees
6. Subtract $\Delta\varepsilon$ from 23.452 294 to find ε.	ε = 23.441 884 degrees
7. If necessary, convert to degrees, minutes and seconds (§ 21).	ε = **23° 26′ 30.8″**

28 Equatorial to ecliptic coordinate conversion

The reverse problem of the previous section is to find the celestial longitude and latitude, λ and β, given the right ascension and declination, α and δ. The formulae are:

$$\lambda = \tan^{-1}\left\{\frac{\sin\alpha\cos\varepsilon + \tan\delta\sin\varepsilon}{\cos\alpha}\right\},$$

$$\beta = \sin^{-1}\{\sin\delta\cos\varepsilon - \cos\delta\sin\varepsilon\sin\alpha\},$$

where ε is the obliquity of the ecliptic (see section 27). These formulae are very nearly identical with those of the previous section with λ, β in place of α, δ and vice versa; the symmetry is not quite complete however as the sign appearing in each formula is reversed.

Consider the example: what are the ecliptic coordinates of a planet whose right ascension and declination are given as $\alpha = 09\text{h}\ 34\text{m}\ 53.6\text{s}$ and $\delta = 19°\ 32'\ 14.2''$? (Once again we use the value of ε for 1980.0)

Method	*Example*
1. Convert α and δ into decimal form (§§ 21 and 7).	$\delta =$ 19.537 278 degrees $\alpha =$ 9.581 556 hours
2. Multiply α by 15 to convert to degrees (§ 22).	$\alpha =$ 143.723 333 degrees
3. Find $\sin\beta = \sin\delta\cos\varepsilon - \cos\delta\sin\varepsilon\sin\alpha$ ($\varepsilon = 23.441\,884$ degrees).	$\sin\beta =$ 0.084 988
4. Take inverse sin to get β in decimal degrees.	$\beta =$ 4.875 306 degrees
5. Find $y = \sin\alpha\cos\varepsilon + \tan\delta\sin\varepsilon$.	$y =$ 0.684 016
6. Find $x = \cos\alpha$.	$x =$ −0.806 169
7. Calculate $\lambda' = \tan^{-1}\left\{\frac{y}{x}\right\}$	$\lambda' =$ −40.313 833 degrees
8. We have to remove the ambiguity which arises here from taking inverse tan. The rule is that λ should lie in the quadrant indicated by the signs of x and y (see Figure 10). Add or subtract 180 or 360 to make this so. If λ' is already in the correct quadrant, $\lambda = \lambda'$.	y positive x negative $\lambda = \lambda' + 180.0$ = 139.686 167 degrees
9. Convert λ and β to minutes and seconds form (§ 21).	$\lambda =$ **139° 41′ 10″** $\beta =$ **4° 52′ 31″**

29 Equatorial to galactic coordinate conversion

Occasionally we need to know the position of a star in relation to the rest of the stars in our Galaxy and to do this we can use the galactic coordinate system. The conversion formulae are:

$$b = \sin^{-1}\{\cos\delta\cos(27.4)\cos(\alpha - 192.25) + \sin\delta\sin(27.4)\},$$

$$l = \tan^{-1}\left\{\frac{\sin\delta - \sin b\sin(27.4)}{\cos\delta\sin(\alpha - 192.25)\cos(27.4)}\right\} + 33.$$

The numbers come from the following facts about our Galaxy: north galactic pole coordinates $\alpha = 192°\,15'$, $\delta = +27°\,24'$; ascending node of galactic plane on equator $l = 33°$.

These are 1950.0 coordinates (see section 33).

The example is: what are the galactic coordinates of a star whose right ascension and declination are $\alpha = 10\text{h}\ 21\text{m}\ 00\text{s}$ and $\delta = 10°\ 03'\ 11''$?

Method	*Example*
1. Convert α, δ into decimal form (§§ 21 and 7).	$\delta =$ 10.053 056 degrees $\alpha =$ 10.350 000 hours
2. Convert α into degrees by multiplying by 15 (§ 22).	$\alpha =$ 155.250 000 degrees
3. Find $\sin b = \cos\delta\cos(27.4)\cos(\alpha - 192.25) + \sin\delta\sin(27.4)$.	$\sin b =$ 0.778 487
4. Take inverse sin to find b in degrees.	$b =$ 51.122 268 degrees
5. Find $y = \sin\delta\text{-}\sin b\sin(27.4)$ and note its sign.	$y =$ −0.183 700 (negative)
6. Find $x = \cos\delta\sin(\alpha - 192.25)\cos(27.4)$ and note its sign.	$x =$ −0.526 097 (negative)
7. Divide y by x.	$y/x =$ 0.349 174
8. Take inverse tan. Now we have to remove the ambiguity which arises from taking the inverse tan. To do so, look at Figure 10 and add or subtract 180 or 360 to bring the result into the correct quadrant, unless it is already in the correct quadrant.	$\tan^{-1}(y/x) =$ 19.247 874 degrees From Fig. 10: +180.0
9. Add 33 to get l.	+ 33.0 $l =$ 232.247 874 degrees
10. Convert l and b into minutes and seconds form (§ 21).	$l =$ **232° 14′ 53″** $b =$ **51° 07′ 20″**

30 Galactic to equatorial coordinate conversion

Given the galactic coordinates, l and b, of a star, what are the corresponding equatorial coordinates, α and δ? To answer this question we need the conversion formulae:

$$\delta = \sin^{-1}\{\cos b \cos(27.4) \sin(l-33) + \sin b \sin(27.4)\},$$

$$\alpha = \tan^{-1}\left\{\frac{\cos b \cos(l-33)}{\sin b \cos(27.4) - \cos b \sin(27.4) \sin(l-33)}\right\}$$

$$+192.25,$$

where both α and δ are expressed in degrees.

As an example we shall find the right ascension and declination of the star whose galactic coordinates are $l = 232° 14' 53''$ and $b = 51° 07' 20''$.

Method	*Example*
1. Convert l and b into decimal form (§ 21).	$b =$ 51.122 222 degrees $l =$ 232.248 056 degrees
2. Find $\sin \delta = \cos b \cos (27.4) \sin (l - 33) + \sin b \sin (27.4)$.	$\sin \delta =$ 0.174 558
3. Take inverse sin to find δ in degrees.	$\delta =$ 10.052 940 degrees
4. Find $y = \cos b \cos (l - 33)$ and note its sign.	$y =$ −0.592 575 (negative)
5. Find $x = \sin b \cos (27.4) - \cos b \sin (27.4) \sin (l - 33)$ and note its sign.	$x =$ 0.786 374 (positive)
6. Divide y by x.	$y/x =$ −0.753 554
7. Take inverse tan. We have to remove the ambiguity which arises from taking the inverse tan. To do so, look at Figure 10 and add or subtract 180 or 360 to bring the result into the correct quadrant, unless it is already there.	$\tan^{-1} (y/x) =$ −36.999 998 (already in correct quadrant)
8. Add 192.25 to find α in degrees.	+ 192.25 $\alpha =$ 155.250 002 degrees
9. Divide by 15 to find α in hours (§ 22).	$\alpha =$ 10.350 000 hours
10. Convert α and δ to minutes and seconds form (§§ 21 and 8).	$\alpha =$ **10h 21m 00s** $\delta =$ **10° 03′ 11″**

31 The angle between two celestial objects

Sometimes it is of interest to know what is the angle between two objects in the sky, and this can be calculated very easily provided their equatorial coordinates (α, δ) or ecliptic coordinates (λ, β) are known. The formulae are:

$$\cos d = \sin \delta_1 \sin \delta_2 + \cos \delta_1 \cos \delta_2 \cos(\alpha_1 - \alpha_2),$$

or

$$\cos d = \sin \beta_1 \sin \beta_2 + \cos \beta_1 \cos \beta_2 \cos(\lambda_1 - \lambda_2),$$

where d is the angle between the objects whose coordinates are α_1, δ_1 (or λ_1, β_1) and α_2, δ_2 (or λ_2, β_2). These formulae are exact and mathematically correct for any values of α, δ or λ, β. However, when d becomes either very small, or close to 180°, your calculator may not have enough precision to return the correct answer, in which case better expressions are

$$d = \sqrt{\cos^2 \delta \cdot \Delta\alpha^2 + \Delta\delta^2}$$

or

$$d = \sqrt{\cos^2 \beta \cdot \Delta\lambda^2 + \Delta\beta^2}$$

where $\Delta\alpha$, $\Delta\delta$ (or $\Delta\lambda$, $\Delta\beta$) are the differences in the two coordinates (i.e. $\Delta\alpha = \alpha_1 - \alpha_2$, etc.). These expressions may be used for values of d within about 10 arcminutes of 0° or 180°. Both $\Delta\alpha$ ($\Delta\lambda$) and $\Delta\delta$ ($\Delta\beta$) must be expressed in the same units (e.g. arcseconds) and d will then be returned in those units.

For example, what is the angular distance between β Orionis ($\alpha = 05\text{h}\ 13\text{m}\ 31.7\text{s}$; $\delta = -8°\ 13'\ 30''$) and α Canis Majoris ($\alpha = 06\text{h}\ 44\text{m}\ 13.4\text{s}$; $\delta = -16°\ 41'\ 11''$)?

Method	*Example*
1. Convert both sets of coordinates to decimal form (§§ 7 and 21).	$\alpha_1 =$ 5.225 472 hours $\delta_1 =$ −8.225 000 degrees $\alpha_2 =$ 6.737 056 hours $\delta_2 =$ −16.686 389 degrees
2. Find $\alpha_1 - \alpha_2$, and convert to degrees by multiplying by 15 (§ 22).	$\alpha_1 - \alpha_2 =$ −1.511 584 hours = −22.673 760 degrees
3. Calculate $\cos d = \sin \delta_1 \sin \delta_2 + \cos \delta_1 \cos \delta_2 \cos (\alpha_1 - \alpha_2)$.	$\cos d =$ 0.915 846
4. Take inverse cos to find d. Convert to minutes and seconds form if required (§ 21).	$d =$ 23.673 859 degrees = **23° 40′ 26″**

32 Rising and setting

During the course of a sidereal day, the stars and other 'fixed' celestial objects appear to move in circles about the rotation axis of the Earth, making one complete revolution in 24 hours. At the moment, there is a star called Polaris very close to the north pole of the Earth's axis so that stars in the northern sky appear to revolve about Polaris. There is nothing special about this star however and no corresponding object exists for the south pole. In any case, the poles are gradually changing their positions in the sky due to precession (see next section) so that Polaris will no longer be the pole star in a few thousand years' time.

The apparent radius of a star's rotation depends, of course, on the angular separation, or *polar distance*, between it and the pole; those stars with a small enough polar distance never dip below the horizon during the course of their rotation. Such stars are called *circumpolar*. As the polar distance increases, however, a point comes when the star just touches the horizon at some time during the day. Stars with polar distances greater than this spend part of their time below the horizon, out of sight to the observer. When the star crosses the horizon on the way down it is said to *set* and as it reappears it is said to *rise*.

The times of rising and setting, and the azimuths at which they occur, can be calculated by the formulae:

$$\mathrm{LST_r} = 24 - \tfrac{1}{15}\cos^{-1}\{-\tan\phi\tan\delta\} + \alpha,$$

$$\mathrm{LST_s} = \tfrac{1}{15}\cos^{-1}\{-\tan\phi\tan\delta\} + \alpha,$$

$$A_\mathrm{s} = 360 - \cos^{-1}\left\{\frac{\sin\delta}{\cos\phi}\right\}, \qquad A_\mathrm{r} = \cos^{-1}\left\{\frac{\sin\delta}{\cos\phi}\right\},$$

where the suffices r and s correspond to rising and setting, A is the azimuth, LST is the local sidereal time in hours, α is the right ascension, δ is the declination, and ϕ is the observer's geographical latitude. These formulae give the times at which the altitude reaches 0°. If your horizon has hills on it, or if you are located on a hill yourself, then your local apparent times of rising and setting will be slightly different. The LST can obviously be converted to GMT and hence to the local civil time by the methods given in sections 15, 13 and 10. Hence, all the circumstances of a star's rising and setting can be calculated. As an example, let us calculate the GMT of rising and setting of a star whose coordinates are $\alpha = 23\text{h}\ 39\text{m}\ 20\text{s}$ and $\delta = 21°\ 42'\ 00''$ on August 24th 1980, and find the corresponding azimuths. The latitude is 30° N and the longitude is 64° E.

Method	*Example*
1. Convert α and δ into decimal form (§§ 7 and 21).	$\delta =$ 21.700 000 degrees $\alpha =$ 23.655 556 hours
2. Find the quantity $\cos A_r = \dfrac{\sin \delta}{\cos \phi}$.	$\cos A_r =$ 0.426 947
3. Take inverse cos* to get A_r.	$A_r =$ 64.726 049 degrees
4. Subtract A_r from 360 to get A_s.	$A_s =$ 295.273 951 degrees
5. Convert to minutes and seconds form (§ 21).	$A_r =$ **64° 43′ 34″** $A_s =$ **295° 16′ 26″**
Now for the times	
6. Find the quantity* $H = \frac{1}{15}\cos^{-1}\{-\tan\phi\tan\delta\}$.	$H =$ 6.885 512 hours
7. Find $\text{LST}_r = 24 + \alpha - H$. If the answer is more than 24, subtract 24.	$\text{LST}_r =$ 16.770 044 hours
8. Find $\text{LST}_s = \alpha + H$. If the answer is more than 24, subtract 24.	$\text{LST}_s =$ 6.541 068 hours
9. Convert these local sidereal times to GST (§ 15).	$\text{GST}_r =$ 12.503 377 hours $\text{GST}_s =$ 2.274 401 hours
10. Convert GST to GMT (§ 13).	$\text{GMT}_r =$ **14h 18m 09s** $\text{GMT}_s =$ **04h 06m 05s**

* If the star's declination is such that it never rises above the horizon, or if it is circumpolar, then you will find that you will be trying to take inverse cos of a number greater than 1 or less than −1. This is impossible and your calculator should respond with 'error'.

Note that the solar times you calculate are appropriate for the date you have used. As here, the setting time may be earlier than the rising time.

Throughout these calculations, we have assumed that the coordinates of the star give its apparent position in the sky. However, there are several effects, including atmospheric refraction (section 34) and parallax (for bodies relatively close to the Earth: section 35), which shift its apparent position and this may alter the apparent times of rising or setting by several minutes. The situation at rising or setting is shown in Figure 11. The celestial body appears to cross the horizon at B, although its 'true' position, as calculated from its uncorrected coordinates, is at A. Provided we know the vertical displacement, x, we can easily calculate the quantities ΔA and y to correct the azimuths and times. The formulae are:

$$y = \sin^{-1}\left\{\frac{\sin x}{\sin \psi}\right\} \text{ degrees}$$

and

$$\Delta A = \sin^{-1}\left\{\frac{\tan x}{\tan \psi}\right\} \text{ degrees,}$$

where

$$\psi = \cos^{-1}\left\{\frac{\sin \phi}{\cos \delta}\right\} \text{ degrees.}$$

Figure 11. The true and apparent positions of a celestial object at rising or setting.

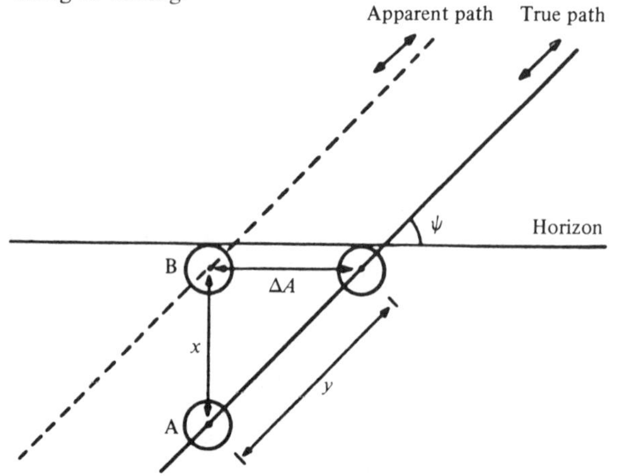

The extra time to be added onto the time of setting, or subtracted from the time of rising, is then given by

$$\Delta t = \frac{240y}{\cos\delta} \text{ seconds,}$$

when y is expressed in degrees.

The value of x due to atmospheric refraction is about 34 minutes of arc. Let us now calculate the effects of refraction on the circumstances of rising and setting in the previous example.

Method	*Example*
1. Calculate $\psi = \cos^{-1}\left\{\frac{\sin\phi}{\cos\delta}\right\}$ (where both ϕ and δ are expressed in decimal degrees).	$\psi =$ 57.443 139 degrees
2. Convert x to decimal degrees (§ 21).	$x =$ 0.566 667 degrees
3. Calculate $\Delta A = \sin^{-1}\left\{\frac{\tan x}{\tan\psi}\right\}$.	$\Delta A =$ 0.361 812 degrees
4. Calculate $A'_r = A_r - \Delta A$ and $A'_s = A_s + \Delta A$, the apparent azimuths of rising and setting. A_r and A_s must be expressed in decimal degrees.	$A'_r =$ 64.364 237 degrees $A'_s =$ 295.635 763 degrees
5. Convert A'_r and A'_s to degrees, minutes and seconds (§ 21).	$A'_r =$ **64° 21′ 51″** $A'_s =$ **295° 38′ 09″**
6. Calculate $y = \sin^{-1}\left\{\frac{\sin x}{\sin\psi}\right\}$.	$y =$ 0.672 321 degrees
7. Calculate $\Delta t = \frac{240y}{\cos\delta}$ (δ expressed in degrees).	$\Delta t =$ 173.664 100 seconds
8. Convert to hours by dividing by 3600.	$\Delta t =$ 0.048 240 hours
9. Find $LST'_r = LST_r - \Delta t$ and $LST'_s = LST_s + \Delta t$, the apparent local sidereal times of rising and setting.	$LST'_r =$ 16.721 804 hours $LST'_s =$ 6.589 308 hours
10. Convert these new times to GST (§ 15), and then to GMT (§ 13).	$GMT'_r =$ **14h 15m 16s** $GMT'_s =$ **04h 08m 59s**

An alternative method of calculating Δt is given in section 34.

33 Precession

In section 18 we found that equatorial coordinates were ideal for fixing the positions of the stars because they were independent of the Earth's motion and therefore constant. This is true to quite a high accuracy, but we find that the coordinates do in fact change slowly with time. This is due to a gyrating motion of the Earth's axis. Rather as the rotation axis of a quickly spinning top revolves slowly about a vertical line, so the rotation axis of the Earth rotates slowly about a fixed direction in space. The motion is called *luni-solar precession* and is caused by the effects of the Moon and Sun on the Earth. We need not be worried by the details. Sufficient to say that the effect is small, the north pole of the Earth making one complete revolution in 25 800 years, but for high accuracy we need to be able to allow for it.

The coordinates α and δ of the stars and galaxies are given in catalogues correct at some particular time or epoch. The ones you are most likely to see will be correct at the epoch 1950.0 (i.e. correct at the beginning of the year 1950) and you can convert the coordinates to the values they will have at some other date using the formulae:*

$$\alpha_1 = \alpha_0 + (3^{s}.073\,27 + 1^{s}.336\,17 \sin \alpha_0 \tan \delta_0) \times N,$$

$$\delta_1 = \delta_0 + (20''.0426 \cos \alpha_0) \times N,$$

where N is the number of years since 1950.0, α_0 and δ_0 are the coordinates at 1950.0, and α_1 and δ_1 are the new coordinates.

To convert from coordinates given at epochs other than 1950.0, use the following formulae with the values of m, n and n' given in Table 5:

$$\alpha_1 = \alpha_0 + (m + n \sin \alpha_0 \tan \delta_0) \times N,$$

$$\delta_1 = \delta_0 + (n' \cos \alpha_0) \times N.$$

* These formulae may not work well for regions around the north and south poles where the magnitude of tan δ tends towards infinity.

Table 5

Epoch	m (seconds)	n (seconds)	n' (arcsec)
1900.0	3.072 34	1.336 46	20.0468
1950.0	3.073 27	1.336 17	20.0426
1975.0	3.073 74	1.336 03	20.0405
2000.0	3.074 20	1.335 89	20.0383

For our example we shall work out the 1979.5 coordinates of a star whose 1950.0 coordinates were $\alpha_0 = 09\text{h}\ 10\text{m}\ 43\text{s}$ and $\delta_0 = 14°\ 23'\ 25''$.

Method	*Example*
1. Convert α_0, δ_0 into decimal form (§§ 21 and 7).	$\alpha_0^{\text{h}} =$ 9.178 611 hours $\delta_0 =$ 14.390 278 degrees
2. Convert α_0 to degrees by multiplying by 15 (§ 22).	$\alpha_0^{\text{d}} =$ 137.679 165 degrees
3. Find $S_1 = (3.073\,27 + 1.336\,17 \sin \alpha_0^d \tan \delta_0) \times N$ (where $N = 1979.5 - 1950.0 = 29.5$).	$S_1 =$ 97.470 657 seconds
4. Divide by 3600 to convert to hours.	$S_1^{\text{h}} =$ 0.027 075 hours
5. Add S_1^{h} to α_0^{h} to get α_1^{h}.	$\alpha_1^{\text{h}} =$ 9.205 686 hours
6. Convert to hours, minutes and seconds (§ 8).	$\alpha_1 =$ **09h 12m 20s**
7. Find $S_2 = (20.0426 \cos \alpha_0^{\text{d}}) \times N$.	$S_2 = -437.167$ 111 arcsec
8. Divide by 3600 to convert to degrees.	$S_2^{\text{d}} =$ −0.121 435 degrees
9. Add S_2^{d} to δ_0 to get δ_1.	$\delta_1 =$ 14.268 843 degrees
10. Convert to degrees, minutes and seconds (§ 21).	$\delta_1 =$ **14° 16′ 08″**

34 Refraction

In all our calculations so far, we have assumed that the light from distant objects reaches us by the most direct route, a straight line. This is not actually the case (except for observations made at the zenith) as the Earth's atmosphere bends the light a little, making the rays reach the ground at a slightly different angle from that which they would have had if the atmosphere were not there (see Figure 12). This is called *atmospheric refraction* and its effect is to make the star appear to be closer to the zenith than it really is. The amount of refraction depends on the *zenith angle* or *zenith distance* (90° − altitude) and on the atmospheric conditions, particularly the temperature and the pressure. For our purposes though it is sufficient to ignore the relatively slight changes from day to day and to adopt a standard index of refraction. Then, if we observe a star with zenith angle ζ from the surface of the Earth, its true zenith angle is given by

$$z = \zeta + R = \zeta + 58''.16 \tan \zeta - 0''.067 \tan^3 \zeta,$$

a formula which is accurate for zenith angles less than about 75°, that is for altitudes *greater* than 15°. The refraction on

Figure 12. Atmospheric refraction.

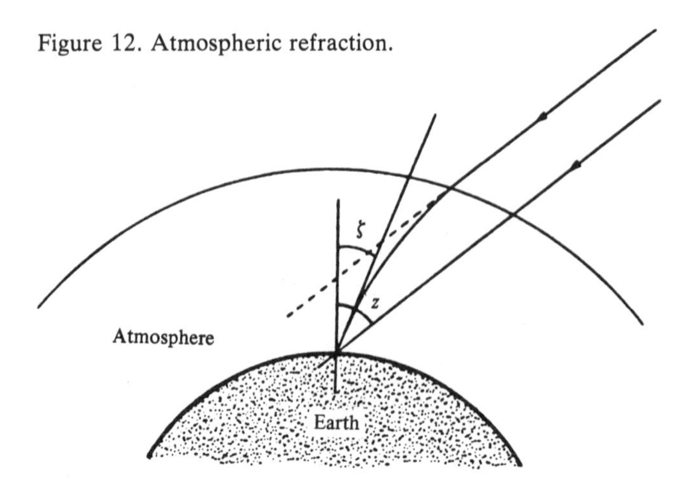

observations made nearer to the horizon than this usually cannot be accurately predicted but is determined by measurement.

The effect of refraction on the right ascension and declination of a star may easily be calculated for zenith angles less than about 45°. If the true coordinates are α and δ, then we will observe apparent coordinates α' and δ' given by

$$\delta' = \delta + 58''.16 \tan \zeta \cos \eta,$$

$$\alpha' = \alpha + 3^s.877 \tan \zeta \frac{\sin \eta}{\cos \delta'},$$

where η is given by

$$\eta = \cos^{-1}\left\{\frac{\sin \phi - \sin \delta \cos z}{\cos \delta \sin z}\right\}.$$

Here, ϕ is the observer's geographical latitude.

We will now calculate the refraction for a star whose equatorial coordinates are $\alpha = 05\text{h}\ 12\text{m}\ 32\text{s}$, $\delta = 45°\ 59'\ 20''$ at a place where its apparent zenith angle is $9°\ 54'\ 16''$. The observer's latitude is 52° N.

Method	*Example*
1. Convert ζ to decimal degrees (§ 21).	$\zeta =$ 9.9044 degrees
2. Calculate $R = 58.16 \tan \zeta - 0.067 \tan^3 \zeta$.	$R =$ 10.15 arcsec
3. Convert R to degrees by dividing by 3600.	$R =$ 0.0028 degrees
4. Find $z = \zeta + R$.	$z =$ 9.9072 degrees
5. Convert δ to decimal degrees (§ 21).	$\delta =$ 45.9889 degrees
6. Calculate $\eta = \cos^{-1}\left\{\frac{\sin \phi - \sin \delta \cos z}{\cos \delta \sin z}\right\}$.	$\eta =$ 48.2951 degrees
7. Calculate $\Delta_1 = 58.16 \tan \zeta \cos \eta$.	$\Delta_1 =$ 6.76 arcsec
8. Convert Δ_1 to degrees by dividing by 3600.	$\Delta_1 =$ 0.0019 degrees
9. Add Δ_1 to δ to find δ'.	$\delta' =$ 45.9908 degrees
10. Convert δ' to degrees, minutes and seconds (§ 21).	$\delta' =$ **45° 59′ 27″**
11. Calculate $\Delta_2 = 3.877 \tan \zeta \frac{\sin \eta}{\cos \delta'}$.	$\Delta_2 =$ 0.727 seconds
12. Add Δ_2 to α to get α'.	$\alpha' =$ **05h 12m 33s**

The magnitude of R right at the horizon is about 34 minutes of arc. Since its effect is to increase the apparent altitude, the times of rising and setting will be earlier and later respectively than they would have been without the atmosphere. The effective length of the day, therefore, is increased by atmospheric refraction. We can calculate its effect on the azimuths and times of rising and setting by the method given in section 32. Alternatively, we can calculate the effect on the hour-angle, H, at rising or setting by

$$\Delta H = \frac{34}{15 \cos \phi \cos \delta \sin H} \text{ minutes of time,}$$

where ΔH is the amount by which the true hour-angle is reduced. Let us apply this formula to the example used in section 32.

Method	*Example*
1. Calculate the magnitude of the hour-angle at rising or setting (the quantity H in § 32). Convert to degrees by multiplying by 15.	$H =$ 6.885 512 hours $=$ 103.282 680 degrees
2. Calculate $\Delta H = \dfrac{34}{15 \cos \phi \cos \delta \sin H}$ ($\phi = 30°$ N; $\delta = 21.7$ degrees).	$\Delta H =$ 2.894 381 minutes
3. Multiply by 60 to convert to seconds; compare this value with Δt calculated in § 32.	$\Delta H =$ **173.66 seconds**

35 Geocentric parallax and the figure of the Earth

In later sections of this book we calculate the coordinates of the Sun, the Moon and other members of our Solar System. These coordinates are the ones which would be observed from the centre of the Earth, called *geocentric coordinates*, and if the celestial body is at a very great distance from the Earth, they are also the coordinates which would be measured by anyone on the Earth's surface. However, objects relatively close at hand like the Sun, and especially the Moon, appear to be at slightly different positions depending upon the exact viewpoint of the observer. This is illustrated in Figure 13 where two observers, O_1 and O_2, are viewing the Moon, M, from the surface of the Earth, E. Each measures the angle between the Moon and a very distant star in the direction S. Since this star is so far away the lines of sight to the star, O_1S and O_2S, are parallel so that both observers see it in the same place in the sky relative to other stars. However, they do not measure the same angles, a_1 and a_2, and hence do not agree about the Moon's apparent position. If a_0 represented, say, the right ascension of the Moon as calculated from the Earth's centre,

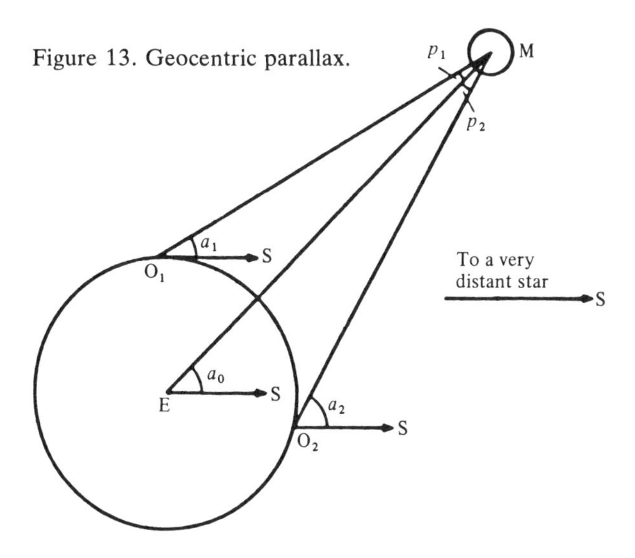

Figure 13. Geocentric parallax.

then each observer would have to add a different correction to a_0 to get a_1 or a_2, the apparent right ascension at each place. This apparent shift of position is known as *geocentric parallax* and we often need to be able to correct for it as, for example, when we wish to calculate the circumstances of an eclipse.

The problem is complicated slightly by the fact that the Earth is not quite spherical, but is, instead, more like a *spheroid of revolution*, being flattened along the line joining the north and south poles. A cross-section through the Earth along any line of longitude would be approximately elliptical, while a cross-section along any line of latitude would be circular. We have to take account of *the figure of the Earth*, its deviation from a perfect sphere, if we are to make precise corrections for parallax. The situation is shown much exaggerated in Figure 14 where the Earth, E, is drawn with its north and south poles, N and S. An observer at O locates his zenith by means of a plumb line to be along the dashed line OZ; the angle this makes with the equator defines his *geographical* or *astronomical latitude*, ϕ. Since the Earth is not quite spherical, his geocentric vertical, EZ′, is slightly different and so too is his *geocentric latitude*, ϕ'. In calculations of the effect of parallax, we need to know the quantities $\rho \sin \phi'$ and

Figure 14. Allowing for the figure of the Earth.

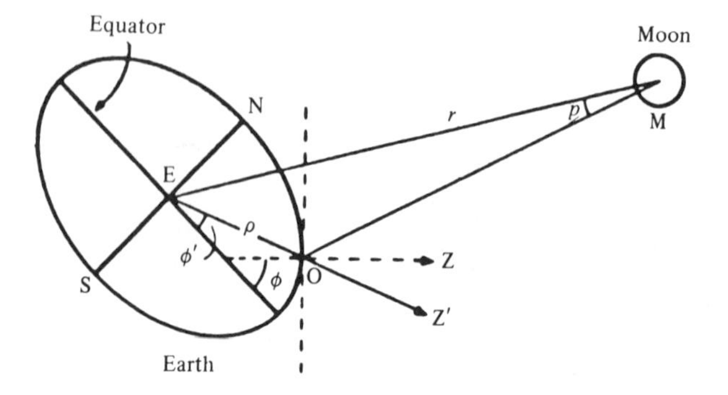

$\rho \cos \phi'$, where ρ is the distance of the observer from the centre of the Earth. For a place whose height above sea-level is h metres, we have

$$\rho \sin \phi' = 0.996\,647 \sin u + \frac{h}{6\,378\,140} \sin \phi,$$

$$\rho \cos \phi' = \cos u + \frac{h}{6\,378\,140} \cos \phi,$$

where

$$u = \tan^{-1}\{0.996\,647 \tan \phi\}.$$

ϕ must be reckoned as positive in the northern hemisphere and negative in the southern hemisphere. For example, let us calculate the values of $\rho \sin \phi'$ and $\rho \cos \phi'$ for an observer whose height above sea-level is 60 metres at longitude 100° W and latitude 50° N.

Method	*Example*	
1. Calculate $u = \tan^{-1}\{0.996\,647 \tan \phi\}$.	$\phi =$	+50.0 degrees
	$u =$	49.905 217 degrees
2. Calculate $h' = (h/6\,378\,140)$.	$h =$	60.0 metres
	$h' =$	0.000 009
3. Calculate $\rho \sin \phi' = 0.996\,647 \sin u + h' \sin \phi$.	$\rho \sin \phi' =$	**0.762 422**
4. Calculate $\rho \cos \phi' = \cos u + h' \cos \phi$.	$\rho \cos \phi' =$	**0.644 060**

In Figure 14, it is the angle p which is formally called the geocentric parallax. This is the angle between the observer and the Earth's centre as seen by the celestial body in question. If the observer views the body right at his horizon (i.e. the zenith angle = 90°), then p is called the *horizontal parallax*. Further, if the observer is also on the equator, this angle becomes the *equatorial horizontal parallax*, and is given the symbol P.

36 Calculating corrections for parallax

If a body has geocentric hour-angle, H, and geocentric right ascension, α, then its apparent hour-angle, H', and right ascension, α' (taking account of parallax), are given by

$$H' = H + \Delta,$$

$$\alpha' = \alpha - \Delta,$$

with

$$\Delta = \tan^{-1}\left\{\frac{\rho \cos \phi' \sin H}{r \cos \delta - \rho \cos \phi' \cos H}\right\},$$

where $\rho \cos \phi'$ is the quantity calculated in section 35 and r is the distance of the body from the centre of the Earth measured in units of Earth-radii, 6378.16 km. If r' is this distance in kilometres, then

$$r = \frac{r'}{6378.16}.$$

r can also be found from the equatorial horizontal parallax of the body, P. Thus

$$r = \frac{1}{\sin P}.$$

The formula for finding the apparent declination, δ', from the geocentric declination, δ, is

$$\delta' = \tan^{-1}\left\{\cos H' \frac{r \sin \delta - \rho \sin \phi'}{r \cos \delta \cos H - \rho \cos \phi'}\right\}.$$

Again, $\rho \sin \phi'$ and $\rho \cos \phi'$ can be found by the method described in section 35.

As an example, let us calculate the apparent right ascension and declination of the Moon on February 26th 1979 at 16h 45m GMT when observed from a location 60 metres above sea-level on longitude 100 °W and latitude 50° N. The geocentric coordinates were $\alpha = 22\text{h}\ 35\text{m}\ 19\text{s}$ and $\delta = -7°\ 41'\ 13''$, and the Moon's equatorial horizontal parallax was $1°\ 01'\ 09''$.

Method	*Example*
1. Convert GMT to GST and hence to LST by the methods of §§ 12 and 14.	GMT = 16h 45m GST = 3.145 733 hours LST = 20.479 066 hours
2. Convert α and δ to decimal form (§§ 7 and 21).	α = 22.588 611 hours δ = −7.686 944 degrees
3. Find the hour-angle, H (§ 24), and convert to degrees (§ 22).	H = −2.109 545 hours = −31.643 175 degrees
4. Find $\rho \cos \phi'$ and $\rho \sin \phi'$ (§ 35).	$\rho \cos \phi'$ = 0.644 060 $\rho \sin \phi'$ = 0.762 422
5. Find $r = (1/\sin P)$ (remember to convert P to decimal degrees first (§ 21)).	P = 1.019 167 degrees r = 56.221 228 Earth-radii
6. Calculate $\Delta = \tan^{-1}\left\{\dfrac{\rho \cos \phi' \sin H}{r \cos \delta - \rho \cos \phi' \cos H}\right\}$.	Δ = −0.350 921 degrees
7. Find $H' = H + \Delta$.	H' = −31.994 096 degrees
8. Convert Δ to hours by dividing by 15 (§ 22).	Δ = −0.023 395 hours
9. Subtract Δ from α to find α'.	α' = 22.612 006 hours
10. Calculate $\delta' = \tan^{-1}\left\{\cos H' \dfrac{r \sin \delta - \rho \sin \phi'}{r \cos \delta \cos H - \rho \cos \phi'}\right\}$.	δ' = −8.538 164 degrees
11. Convert α' and δ' to minutes and second form (§§ 8 and 21).	α' = **22h 36m 43s** δ' = **−8° 32′ 17″**

Such lengthy calculations are strictly only necessary for the Moon which has a very large parallax. The Sun, planets and comets usually have much smaller values, enabling us to simplify the formulae slightly without serious loss of accuracy. Let r again denote the distance of the body from the centre of the Earth, but this time measured in astronomical units (AU). Then

$$\pi = \frac{8.794}{r} \text{ arcseconds}$$

and

$$\alpha' = \alpha - \frac{\pi \sin H \rho \cos \phi'}{\cos \delta},$$

$$\delta' = \delta - \pi(\rho \sin \phi' \cos \delta - \rho \cos \phi' \cos H \sin \delta).$$

π is the symbol often used for parallax. Take care to distinguish its use for parallax from its use to represent the circular constant 3.141 592 654.

Let us now calculate the apparent position of the Sun when observed by the same observer at the same time as in the previous example. The geocentric right ascension of the Sun was 22h 36m 44s and its declination was $-8° 44' 24''$. Its distance was 0.9901 AU.

Method	*Example*
1. Calculate $\pi = \frac{8.794}{r}$.	$\pi =$ 8.881 931 arcsec
2. Convert to degrees by dividing by 3600.	$\pi =$ 0.002 467 degrees
3. Convert to hours by dividing by 15 (§ 22).	$\pi =$ 0.000 164 hours
4. Convert GMT to GST and hence to LST by the methods of §§ 12 and 14.	GST = 3.145 733 hours LST = 20.479 066 hours
5. Convert α and δ to decimal form (§§ 7 and 21)	$\alpha =$ 22.612 222 hours $\delta =$ −8.740 000 degrees
6. Find $\rho \cos \phi'$ and $\rho \sin \phi'$ (§ 35).	$\rho \cos \phi' =$ 0.644 060 $\rho \sin \phi' =$ 0.762 422
7. Find the hour-angle, H, and convert to degrees (§§ 24 and 22).	$H =$ −2.133 156 hours = −31.997 340 degrees
8. Calculate $\Delta_1 = \frac{\pi \sin H \rho \cos \phi'}{\cos \delta}$ (π expressed in hours).	$\Delta_1 =$ −0.000 057 hours
9. Subtract from α to get α'.	$\alpha' =$ 22.612 279 hours
10. Find $\Delta_2 = \pi(\rho \sin \phi' \cos \delta - \rho \cos \phi' \cos H \sin \delta)$ (π expressed in degrees).	$\Delta_2 =$ 0.002 064 degrees
11. Subtract from δ to get δ'.	$\delta' =$ −8.742 064 degrees
12. Convert α' and δ' to minutes and seconds form (§§ 8 and 21).	$\alpha' =$ **22h 36m 44s** $\delta' =$ **−8° 44′ 31″**

Note that the correction for parallax has had hardly any effect in this case. Except for the Moon, geocentric parallax can often be ignored. Note also that the Sun and Moon have almost the same apparent positions in this example; we have chosen the moment of a total solar eclipse (see section 70).

37 Heliographic coordinates

Heliographic coordinates enable us to define the position of any point (such as a sunspot) on the surface of the Sun. As with any other set of astronomical coordinates, latitudes are referred to a fundamental plane and longitudes to a fixed point in that plane. In this case the fundamental plane is taken to be the solar equator, inclined at an angle $I = 7° 15'$ to the ecliptic, and the fixed point is the present position of the point occupied by the ascending node of the solar equator on the ecliptic at noon on January 1st 1854. There are no permanent features on the Sun's disc by which we can locate this point, so we have to work out its position assuming a rotation period of 25.38 days.

The situation is illustrated in Figure 15. The sphere represents the surface of the Sun and the great circle ONJ the solar equator. $P_N C$ is the rotation axis of the Sun and any point on the equator rotates in the direction from O to N. The plane of the ecliptic intersects the Sun's surface along the great circle ♈EN; the point N is therefore the ascending node of the solar

Figure 15. Defining heliographic coordinates.

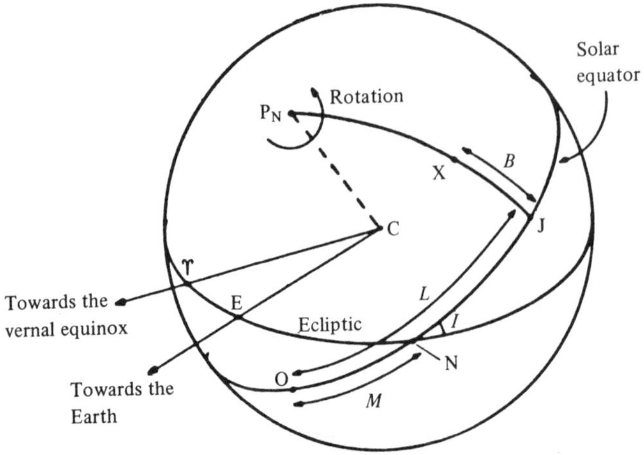

equator on the plane of the ecliptic. Imaginary lines drawn from the centre of the Sun, C, towards the Earth and towards the vernal equinox cut the Sun's surface at E and ♈ respectively. O is the point which, at midday on January 1st 1854, was at N. A sunspot at X has heliographic latitude B (positive north of the equator, negative south of it) and heliographic longitude L, reckoned in the same sense as the solar rotation and measured along the equator from O.

When we observe the Sun (which must only be by projection onto a screen) we see a flat disc, the centre of which is the point E. This is shown in Figure 16, together with the point P_N, the north pole of the Sun's rotation axis, and X, the position of the sunspot. The dashed line N′S′ represents the projection of the Earth's axis of rotation, NS, onto the disc. We define the position of X by the coordinates ρ_1 and θ. The trick is to turn ρ_1 and θ into B and L.

To do this we need first to calculate the heliographic coordinates, B_0 and L_0, of the centre of the disc, E. The equations are:

Figure 16. The Sun's disc.

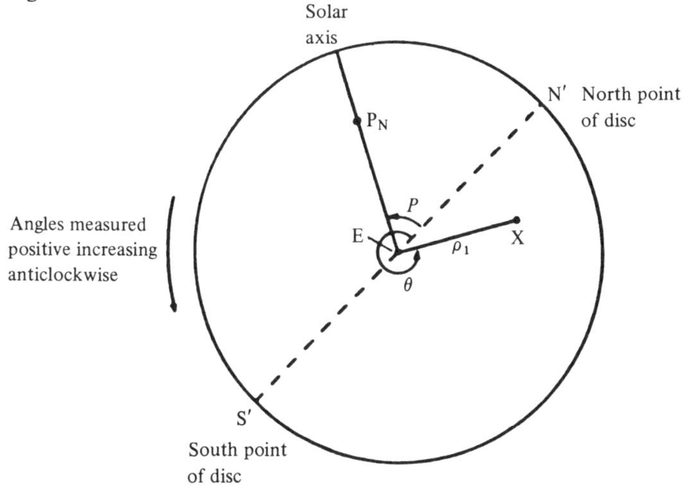

$$L_0 = \tan^{-1}\left\{\frac{\sin(\Omega - \lambda_\odot)\cos I}{-\cos(\Omega - \lambda_\odot)}\right\} + M,$$

$$B_0 = \sin^{-1}\{\sin(\Omega - \lambda_\odot)\sin I\},$$

where $\lambda_\odot$ is the geocentric ecliptic longitude of the Sun, I is the inclination of the solar equator to the plane of the ecliptic ($= 7° 15'$), Ω is the longitude of the ascending node (the angle ♈N in Figure 15), and M is the angle between O and N (Figure 15). Ω is given by

$$\Omega = 74° 22' + 84'T,$$

where T is the number of Julian centuries since the epoch 1900 January 0.5. M is given by

$$M = 360 - M',$$

with

$$M' = \frac{360}{25.38}(\mathrm{JD} - 2\,398\,220.0),$$

where JD is the Julian date. M' must be reduced to the range 0–360 by subtracting integral multiples of 360.

For example, let us calculate the heliographic coordinates of the centre of the solar disc on May 1st 1979. The geocentric longitude of the Sun, $\lambda_\odot$, can be found by the method given in section 42; its value on this day was 40° 02′ 44″.

Method	*Example*
1. Find the Julian date (§ 4).	JD = 2 443 994.50
2. Subtract 2 415 020 and divide by 36 525 to find T in centuries since 1900 January 0.5.	$T =$ 0.793 279 centuries
3. Calculate $\Delta = 84T$. Divide by 60 to convert to degrees.	$\Delta =$ 66.635 400 arcmin = 1.110 590 degrees
4. Convert 74° 22′ to decimal degrees (§ 21) and add to Δ to find ☊.	☊ = 75.477 257 degrees
5. Convert $\lambda_\odot$ to decimal degrees (§ 21). (You can calculate $\lambda_\odot$ by the method of § 42.)	$\lambda_\odot =$ 40.045 556 degrees
6. Find $y = \sin(☊ - \lambda_\odot)\cos I$ ($I = 7.25$ degrees).	$☊ - \lambda_\odot =$ 35.431 701 degrees $y =$ 0.575 097
7. Find $x = -\cos(☊ - \lambda_\odot)$.	$x =$ −0.814 807
8. Find $A' = \tan^{-1}\{y/x\}$. We have to remove the ambiguity introduced by taking inverse tan. To do so look up Figure 10 and add or subtract 180 or 360 until A' is in the correct quadrant. If it is already in the correct quadrant, $A = A'$.	$A' =$ −35.214 735 y positive x negative +180.0 $\therefore A =$ 144.785 265
9. Calculate $M' = \frac{360}{25.38}(\text{JD} - 2\,398\,220)$. Subtract multiples of 360 to bring it back into the range 0–360.	$M' =$ 649 283.687 9 −360 × 1803 = 203.687 9 degrees
10. Find $M = 360 - M'$.	$M =$ 156.312 1 degrees
11. Add M to A to find L_0. Subtract 360 if more than 360.	$L_0 =$ **301.097 degrees**
12. Calculate $B_0 = \sin^{-1}\{\sin(☊ - \lambda_\odot)\sin I\}$.	$B_0 =$ **4.196 degrees**

The *Astronomical Ephemeris* lists these values as $L_0 =$ 301.09 degrees and $B_0 = 4.20$ degrees.

In addition to B_0 and L_0 we also need the position-angle of the Sun's rotation axis, the angle P in Figure 16. This is given by

$$P = \theta_1 + \theta_2,$$

with

$$\theta_1 = \tan^{-1}\{-\cos \lambda_\odot \tan \varepsilon\}$$

and

$$\theta_2 = \tan^{-1}\{-\cos (\Omega - \lambda_\odot) \tan I\},$$

where ε is the obliquity of the ecliptic (see section 27).

For example, what was the value of the P on May 1st 1979?

Method	*Example*
Referring to the previous example for values of $\lambda_\odot$, Ω and I:	$\lambda_\odot =$ 40.045 556 degrees
	$\Omega =$ 75.477 257 degrees
	$I =$ 7.25 degrees
1. Calculate $\theta_1 = \tan^{-1}\{-\cos \lambda_\odot \tan \varepsilon\}$ (with $\varepsilon = 23.442$ degrees).	$\theta_1 = -18.363\ 185$ degrees
2. Calculate $\theta_2 = \tan^{-1}\{-\cos (\Omega - \lambda_\odot) \tan I\}$.	$\theta_2 = -5.917\ 948$ degrees
3. Find $P = \theta_1 + \theta_2$.	$P =$ **−24.281 133 degrees**

The *Astronomical Ephemeris* gives $P = -24.28$ degrees.

We are now in a position to calculate the heliographic coordinates of the sunspot X, given its position-angle, θ (see Figure 16), and ρ_1, the angle subtended at the Earth by X and E. The formulae are as follows:

$$B = \sin^{-1}\{\sin B_0 \cos \rho + \cos B_0 \sin \rho \cos (P - \theta)\},$$

$$L = \sin^{-1}\left\{\frac{\sin \rho \sin (P - \theta)}{\cos B}\right\} + L_0,$$

with

$$\rho = \sin^{-1}\left\{\frac{\rho_1}{S}\right\} - \rho_1,$$

where S is the angular radius of the Sun.

Continuing our example, what were the heliographic coordinates of a sunspot measured at position-angle $\theta = 220°$ and displacement $\rho_1 = 10.5$ arcminutes on May 1st 1979? The angular radius of the Sun was 15′ 54″.

Method	*Example*	
1. Calculate L_0, B_0 and P (see previous examples).	$L_0 =$	301.097 degrees
	$B_0 =$	4.196 degrees
	$P =$	−24.281 degrees
2. Find $\sin^{-1}\left\{\frac{\rho_1}{S}\right\}$	$S =$	15.9 arcmin
(remember to convert S to decimal arcminutes first).	$\sin^{-1}\left\{\frac{\rho_1}{S}\right\} =$	41.328 659 degrees
3. Convert ρ_1 to decimal degrees (§ 21) and subtract to find ρ.	$\rho_1 =$	0.175 000 degrees
	$\therefore \rho =$	41.153 659 degrees
4. Calculate $B = \sin^{-1}\{\sin B_0 \cos \rho + \cos B_0 \sin \rho \cos (P - \theta)\}$.	$(P - \theta) =$	−244.281 degrees
	$B =$	**−13.280 690 degrees**
5. Calculate $A = \sin^{-1}\left\{\frac{\sin \rho \sin (P - \theta)}{\cos B}\right\}$.	$A =$	37.530 005 degrees
6. Add L_0 to find L; subtract 360 if L is greater than 360.	$L =$	**338.627 005 degrees**

38 Carrington rotation numbers

Solar rotations are numbered by the Carrington Rotation Number, CRN, the first of which began on November 9th 1853. One rotation is the period during which the value of L_0 (section 37) decreases by 360°, and its mean length is 27.2753 days. We can calculate CRN quite accurately by noting from the *Astronomical Ephemeris* that rotation number 1690 began on 1979 December 27.84. Thus

$$\mathrm{CRN} = 1690 + \left[\frac{\mathrm{JD} - 2\,444\,235.34}{27.2753}\right],$$

where JD is the Julian date. Round the result to the nearest integer. You may be in error by ±1 just at the point where the rotation number changes.

Example. What was the CRN on January 27th 1975?

Method	*Example*
1. Calculate the Julian date (§ 4).	JD = 2 442 439.50
2. Find $\mathrm{CRN} = 1690 + \left[\frac{\mathrm{JD} - 2\,444\,235.34}{27.2753}\right]$, rounding the result to the nearest integer.	CRN = **1624**

39 Atmospheric extinction

The light which reaches us on the surface of the Earth from heavenly bodies has to pass first through the atmosphere where some of it is scattered by dust, electrons, oxygen and nitrogen molecules, and other sundry particles. The amount of scattering depends on the physical conditions in the atmosphere (it will be enhanced, for example, by extra dust from a volcanic eruption) and on the wavelength of the light. In general, the shorter wavelengths (blue) are scattered much more than the longer wavelengths (red); for this reason, the sky looks blue (we see the scattered light) and the apparent colour of a star observed from the Earth's surface is reddened. If we take the visual wavelengths as a whole, we can make a rough estimate of the amount of absorption to expect when the atmosphere is clear, from

$$\Delta m = \frac{0.2}{\cos z} \text{ magnitudes,}$$

where Δm is the quantity to be added to the magnitude, and z is the zenith angle ($z = 90° -$ altitude). For example, a planet whose altitude is 15° may appear dimmer by about 0.8 magnitudes in good conditions when the atmosphere is clear; in general this will be an underestimate. The formula breaks down for zenith angles greater than about 85°.

The Sun

The nearest star to the Earth is the Sun, being some 91 million miles distant at its closest approach. The sunlight reaching us is already eight minutes old when we see it, having taken this long to travel the radius of the Earth's orbit. Yet despite this distance the Sun is so huge that it appears as one of the largest celestial objects in the sky, equalled only by the Moon which, by coincidence, has more or less the same angular size, and it is certainly the brightest. It dominates the Solar System, controlling the motions of the planets and supplying the energy needed for life on Earth. Although we always know by experience approximately where the Sun is in the sky, we often need to know its position more accurately as, for example, when we wish to calculate an eclipse or the orientation of a sundial. The next few sections deal with methods for calculating the Sun's orbit, distance from the Earth, apparent angular size, the times of sunrise and sunset, the solar elongations of other celestial bodies, the equation of time which you will need if you wish to set your watch by your sundial, and the duration of twilight.

40 Orbits

The motions of the planets around the Sun, and of the satellites about their planets, are all controlled by the action of gravity, that is by the mutual force of attraction between the masses. One of the consequences of the way this force varies with distance is that the planetary orbits trace out the forms of ellipses (Figure 17), geometrical shapes with well-known mathematical properties which enable us to calculate a planet's course precisely. You can imagine an ellipse as a squashed circle; in fact, a circle is a special case of an ellipse where the two *foci*, S and S′, have moved together into the middle. The amount of squashing is measured by the *eccentricity*, e; for a circle, $e = 0$, and the most flattened ellipses have values of e approaching 1. Most planetary orbits have eccentricities less than 0.1 so that their deviations from circular orbits are small. This is fortunate as it enables us to calculate planetary positions quite accurately by relatively simple methods.

All of the planetary orbits have the Sun at one of the foci, S. In Figure 17 the planet V moves in the direction of the arrow around the ellipse, its distance from the Sun varying from a minimum at A to a maximum at B. These points are called *perihelion* and *aphelion* respectively. The line joining the

Figure 17. An orbital ellipse.

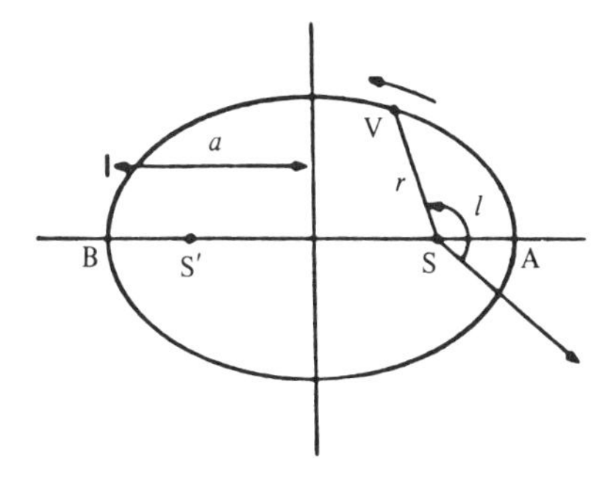

planet to the Sun, *r*, is called the radius vector, and the angle, *l*, it makes with a fixed direction in space defines the position of the planet in its orbit at any time. The size of the ellipse is defined by the *semi-major axis, a*.

41 The apparent orbit of the Sun

During the course of a year, the Earth moves in its own elliptical orbit around the Sun, making one complete revolution in about $365\frac{1}{4}$ days. Viewed from the Earth, it seems to us that the Sun moves in orbit around the Earth and for the purposes of calculating the Sun's position it is convenient to regard this as the case. Hence we now assume that it is we who are at the focus and the Sun describes an ellipse about us. When the Sun is closest to the Earth, we say it is at *perigee* and when it is farthest away it is at *apogee*.

Since the plane which contains the Sun–Earth orbit defines the plane of the ecliptic, it is particularly easy to calculate the Sun's apparent motion as we do not have to worry about deviations from the ecliptic. Once we have calculated the ecliptic longitude, we have defined the Sun's position as the ecliptic latitude is zero.

42 Calculating the position of the Sun

The first thing to do is to define the epoch on which we shall base our calculation; we choose 1980 January 0.0. From the *Astronomical Ephemeris* we find that the Sun's mean ecliptic longitude at the epoch was $\varepsilon_g = 278.833\,540$ degrees; this is the position it would have had if it had been moving in a circular orbit rather than an ellipse. The value of ε_g represents our starting point. We simply have to add on the correct number of degrees moved by the Sun since then, and to make due allowance for its elliptical motion, to find where it is at any other time. To do so we need two other constants: $\varpi_g = 282.596\,403$, the longitude of the Sun at perigee, and $e = 0.016\,718$, the eccentricity of the Sun–Earth orbit.

We now imagine that the Sun moves in a circle around the Earth at a constant speed, rather than along the ellipse which it actually traces. We can easily calculate the angle, M, through which this ficticious *mean Sun* has moved since it passed through perigee:

$$M = \frac{360}{365.2422} d \text{ degrees},$$

where d is the number of days since perigee, because during the course of one tropical year of 365.2422 days the Sun completes a circle of 360°. M is called the *mean anomaly*. But rather than basing our calculations on the moment of perigee, we have decided to use the epoch 1980.0. Then if D is the number of days since the epoch, M is given by (Figure 18):

$$M = \frac{360}{365.2422} D + \varepsilon_g - \varpi_g$$

where ε_g and ϖ_g are the mean longitudes of the Sun at the epoch and perigee respectively (values given in Table 6).

M refers to the motion of a mean Sun moving in a circle. We actually need the *true anomaly*, ν, which applies for the true

Figure 18. Defining the orbit of the Sun.

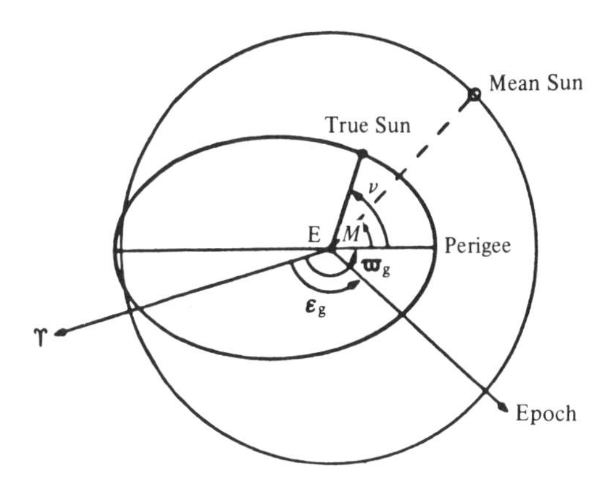

Table 6. Details of the Sun's apparent orbit

ε_g (ecliptic longitude at epoch 1980.0)	$= 278.833\,540$ degrees
ϖ_g (ecliptic longitude of perigee)	$= 282.596\,403$ degrees
e (eccentricity of orbit)	$= 0.016\,718$
r_0 (semi-major axis)	$= 1.495\,985 \times 10^8$ km
θ_0 (angular diameter at $r = r_0$)	$= 0.533\,128$ degrees

motion of the Sun in an ellipse. This can be found from the *equation of the centre*, which is (to a sufficient accuracy for our purposes; see Figure 25):

$$\nu = M + \frac{360}{\pi} e \sin M,$$

where ν and M are expressed in degrees and $\pi = 3.141\,592\,7$. Having found ν, we simply add ϖ_g (Figure 18) to get the longitude of the Sun, $\lambda_\odot$. Hence

$$\lambda_\odot = \nu + \varpi_g,$$

or

$$\lambda_\odot = \frac{360}{365.2422} D + \frac{360}{\pi} e \sin \left\{ \frac{360}{365.2422} D + \varepsilon_g - \varpi_g \right\} + \varepsilon_g.$$

Having calculated $\lambda_\odot$, the method given in section 27 may be used to find the right ascension and declination (remembering that the ecliptic latitude is zero).

Let us clarify all this by working out an example: what were the right ascension and declination of the Sun on July 27th 1980?

Method	*Example*
1. Find the number of days since January 0.0 at the beginning of the year (§ 3).	July 27th = 182 + 27 = 209
2. Add 365 days for every year since 1980 plus 1 extra day for every leap year (see Table 2). The result is D.	+ 0 $\therefore D = 209$ days
3. Calculate $N = \frac{360}{365.2422} D$; subtract (or add) multiples of 360 until N lies in the range 0–360.	$N = 206.000\,292$ degrees
4. Find $M = N + \varepsilon_g - \varpi_g$ (Table 6). If the result is negative, add 360.	$M = 202.237\,429$ degrees
5. Find $E_c = \frac{360}{\pi} e \sin M$ ($\pi = 3.141\,592\,7$ and e from Table 6).	$E_c = -0.725\,004$ degrees
6. Find $\lambda_\odot = N + E_c + \varepsilon_g$. If the result is more than 360, subtract 360. This is the Sun's geocentric ecliptic longitude.	$(\nu = M + E_c)$ $\lambda_\odot = 124.108\,828$ degrees
7. Now convert to right ascension and declination (§ 27). Remember that $\beta_\odot = 0$.	$\alpha_\odot =$ **08h 25m 44s** $\delta_\odot =$ **19° 13′ 53″**

The *Astronomical Ephemeris* gives $\alpha = 08^h\,25^m\,44^s$ and $\delta = 19°\,13'\,46''$ so our result is really quite accurate. In general we should find that we can calculate α to within about 10s and δ to within a few minutes of arc by this method. The inaccuracies arise because we have only used the first term in the equation of the centre, we have not taken account of all sorts of tiny perturbations due to the influences of the other planets in the Solar System, and because we have made no allowances for the various corrections like parallax, refraction, precession, etc. In this particular example, the corrections in right ascension seem to have cancelled themselves since we get the same result as that given in the *Astronomical Ephemeris*.

43 Calculating orbits more precisely

In this section we discover how to find the true anomaly, ν, by a slightly more accurate method.* For most purposes the accuracy of the simpler method given in section 42 will suffice, but if you have a good programmable calculator, or even access to a digital computer, you may find this section useful.

As before, we find the mean anomaly, M, by

$$M = \frac{360}{365.2422} D + \varepsilon_g - \varpi_g$$

where M, ε_g and ϖ_g are all expressed in degrees. Then we find the *eccentric anomaly*, E, which is defined on the diagram in Figure 19. E is given by Kepler's equation, named after the famous astronomer, Johannes Kepler, who made a detailed study of the planets:

$$E - e \sin E = M,$$

where this time E and M are expressed in circular measure, or *radians*. Unfortunately, this equation is not easily solved, but with the aid of an iterative routine a numerical solution may be reached. Such a routine is listed in R2 and is appropriate for values of e less than about 0.1 (but see section 57 for larger values of e).

Having found a solution to Kepler's equation, we can find the true anomaly, ν, from

$$\tan \frac{\nu}{2} = \left[\frac{1+e}{1-e}\right]^{\frac{1}{2}} \tan \frac{E}{2},$$

where the angles are all expressed in *radians*, and then, as in the previous section, the ecliptic longitude is given by

$$\lambda_{\odot} = \nu + \varpi_g \text{ degrees}$$

(remember to convert ν from radians to degrees before adding ϖ_g).

* Figure 25 (section 52) shows the error introduced by using only the first term in the equation of the centre.

Routine R2. To find a solution to Kepler's equation $E - e \sin E = M$ for $e \lesssim 0.1$. All angles expressed in radians.

1. First guess put $E = E_0 = M$.
2. Find the value of $\delta = E - e \sin E - M$.
3. If $|\delta| \leqslant \varepsilon$ go to step 6.
   If $|\delta| > \varepsilon$ proceed with step 4.
   ε is the required accuracy ($=10^{-6}$ radians).
4. Find $\Delta E = \dfrac{\delta}{1 - e \cos E}$.
5. Take new value $E_1 = E - \Delta E$. Go to step 2.
6. The present value of E is the solution, correct to within ε of the true value.

Figure 19. True and eccentric anomalies.

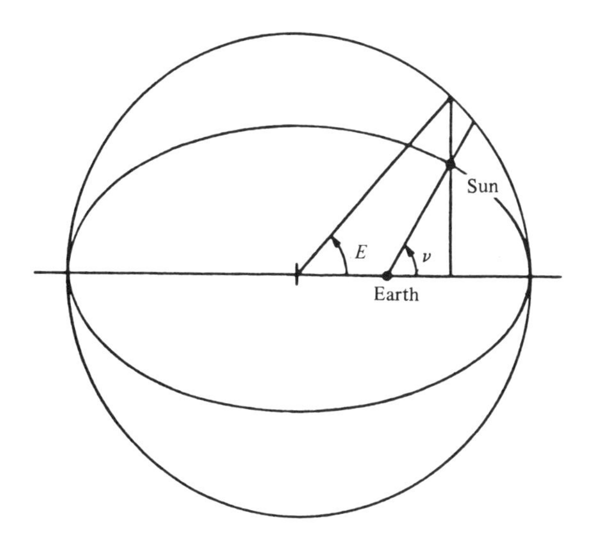

Let us use this method to solve another example: what were the right ascension and declination of the Sun on July 27th 1978?

Method	*Example*
1. Find the number of days since the beginning the year in question (§ 3).	July 27th = 208 days
2. Add 365 days for every year since 1980 plus 1 extra day for every leap year. (In this case we subtract since 1978 was before 1980.) The total is D.	−365 (1979) −365 (1978) $D = -522$ days
3. Calculate $N = \dfrac{360}{365.2422}D$. Add or subtract multiples of 360 until N lies in the range 0–360.	$N = -514.507\,907$ degrees $= 205.492\,093$ degrees
4. Find $M = N + \varepsilon_g - \varpi_g$ (Table 6). If the result is negative, add 360.	$M =$ 201.729 230 degrees
5. Convert M to radians by multiplying by $\dfrac{\pi}{180}$ ($\pi = 3.141\,592\,7$).	$M =$ 3.520 839 radians
6. Use routine R2 to find a solution for E.	$E =$ 3.514 745 radians
7. Find $\tan\dfrac{\nu}{2} = \left[\dfrac{1+e}{1-e}\right]^{\frac{1}{2}} \tan\dfrac{E}{2}$ (remember that E and ν are in radians).	$\tan\dfrac{\nu}{2} =$ −5.386 726
8. Take inverse tan to find $\dfrac{\nu}{2}$ (in radians).	$\dfrac{\nu}{2} =$ −1.387 244 radians
9. Multiply by 2 and convert to degrees by multiplying by $\dfrac{180}{\pi}$.	$\nu = -158.966\,491$ degrees
10. Find $\lambda_\odot = \nu + \varpi_g$. If the answer is more than 360, subtract 360. If it is negative, add 360. This is the Sun's ecliptic longitude.	$\lambda_\odot =$ 123.629 912 degrees
11. Convert $\lambda_\odot$ to right ascension and declination (§ 27) remembering that $\beta_\odot = 0$.	$\alpha_\odot =$ **08h 23m 46s** $\delta_\odot =$ **19° 20′ 38″**

The *Astronomical Ephemeris* gave $\alpha = 08\text{h}\ 23\text{m}\ 44\text{s}$ and $\delta = 19°\ 20'\ 38''$.

44 Calculating the Sun's distance and angular size

Having found the true anomaly, ν, by the methods of sections 42 or 43, we can easily calculate the Sun–Earth distance, r, and the Sun's angular size (i.e. its angular diameter), θ. The formulae are:

$$r = \frac{r_0(1-e^2)}{(1+e\cos\nu)},$$

$$\theta = \theta_0 \frac{(1+e\cos\nu)}{(1-e^2)},$$

where r_0 is the semi-major axis, θ_0 is the angular diameter when $r = r_0$, and e is the eccentricity of the orbit. These constants are given in Table 6. Continuing the example of section 42 we can find r and θ for the Sun on July 27th 1980.

Method	*Example*
1. Find the true anomaly, ν (§§ 42 or 43).	$\nu = 201.512\,425$ degrees
2. Find $f = \frac{1+e\cos\nu}{(1-e^2)}$ (Table 6 for e).	$f =$ 0.984 722
3. Then $r = \frac{r_0}{f}$ (Table 6 for r_0).	$r =$ **1.519 196 × 10⁸ km**
4. And $\theta = f\theta_0$ (Table 6 for θ_0).	$\theta =$ 0.524 957 degrees
5. Convert θ to minutes and seconds form (§ 21).	$\theta =$ **0° 31′ 30″**

The *Astronomical Ephemeris* gives $\theta = 0°\,31'\,33''$ and in general we should be within a few seconds of arc of the correct value. It is interesting to note that the Sun's light took r/c seconds to reach us, where $c = 3 \times 10^5$ km s^{-1}. In this case the light travel time was 506 seconds, during which interval the Sun moved about 21 seconds of arc. Strictly speaking, we should subtract this from the calculated position to find the Sun's apparent position.

45 Sunrise and sunset

In section 32 we found how to calculate the rising and setting time of any celestial object whose equatorial coordinates were known. We have calculated the right ascension and declination of the Sun (sections 42 and 43) so that we can apply the same method to find the times of sunrise and sunset. The problem is complicated, however, by the fact that the Sun is in continual motion along the ecliptic and its equatorial coordinates are therefore changing. The values of α and δ we have calculated are correct only for the time we have chosen. (In the example of section 42, this time was the midnight between July 26th and July 27th. By the time the Sun had risen the next morning it had moved about a quarter of a degree from its midnight position, and by sunset about three quarters of a degree). Provided that we do not require high accuracy in our calculations we can ignore the Sun's motion and simply take the midnight α and δ as correct for sunrise and sunset. The results are then within a few minutes of their correct values.

We can do better, however, by calculating the Sun's coordinates for the two midnights straddling the required sunrise and sunset, and perform an interpolation between them. (We do a similar thing for the Moon in section 66.) The formula is

$$T = \frac{24.07 \times \mathrm{ST1}}{24.07 + \mathrm{ST1} - \mathrm{ST2}} \text{ hours}$$

where T is the actual sidereal time of rising or setting, ST1 is the sidereal time of rising or setting calculated from α_1 and δ_1 appropriate to the midnight before, and ST2 is the sidereal time of rising or setting calculated from α_2 and δ_2 appropriate to the midnight after. It is not really necessary to perform all the calculations of sections 42 or 43 twice to find the two sets of coordinates. Rather, it will suffice to calculate α_1 and δ_1 for the preceding midnight, and then to note that during 24 hours the Sun's ecliptic longitude increases by about 0.985 647 degrees. Add this to $\lambda_\odot$ and hence find α_2 and δ_2.

Further refinements include taking account of refraction by the Earth's atmosphere (section 34) and geocentric parallax (section 35). We must also consider the finite diameter of the Sun's disc: times of sunrise and sunset are usually quoted as those appropriate to the upper limb.

For example, let us calculate the times of sunrise and sunset (upper limb) over a level horizon at sea-level on September 7th 1979, as observed from a place at longitude 0° and latitude 52° N. We shall take the Sun's angular diameter to be 0.533 degrees, its horizontal parallax to be 8.79 arcseconds and the refraction due to the atmosphere as 34 minutes of arc.

Method	*Example*
1. Calculate the coordinates of the Sun's centre by §§ 42 or 43.	$\lambda_\odot = 163.778\,867$ degrees $\alpha_1 = 11.003\,687$ hours $\delta_1 = 6.380\,389$ degrees
2. Add 0.985 647 to $\lambda_\odot$, and recalculate α and δ from the new value (§ 27).	$\lambda_2 = 164.764\,514$ degrees $\alpha_2 = 11.064\,683$ hours $\delta_2 = 6.000\,751$ degrees
3. Calculate the local sidereal times of rising and of setting appropriate to those coordinates (§ 32).	$ST1_r = 4.455\,106$ hours $ST1_s = 17.552\,268$ hours $ST2_r = 4.549\,199$ hours $ST2_s = 17.580\,167$ hours
4. Apply the formula $T = \dfrac{24.07 \times ST1}{24.07 + ST1 - ST2}$ to the pairs of values for rising and setting.	$T_r = 4.472\,590$ hours $T_s = 17.572\,636$ hours
5. Calculate the correction, Δt, due to parallax, refraction, and the Sun's finite diameter (§ 32). Use the average value of δ_1 and δ_2 $\left(= \dfrac{\delta_1 + \delta_2}{2}\right)$.	$\delta' = \dfrac{\delta_1 + \delta_2}{2} = 6.190\,570$ degrees $\psi = 37.567\,754$ degrees $x = \dfrac{0^\circ.533}{2} + 8''.79 + 34'$ $= 0.835\,608$ degrees $y = 1.370\,608$ degrees $\Delta t = 330.88$ seconds
6. Convert Δt to hours by dividing by 3600; add the result to T_s and subtract it from T_r.	$\Delta t = 0.091\,910$ hours $T_r = 4.380\,680$ hours $T_s = 17.664\,546$ hours
7. Convert T_r and T_s to GMT (§ 13). (Note that in this case the longitude is 0° so that the local sidereal times are Greenwich sidereal times.)	$GMT_r =$ **05h 20m** $GMT_s =$ **18h 35m**

46 Twilight

Whenever the Sun is less than 18° below the horizon, after sunset or before sunrise, the light scattered by the upper atmosphere illuminates the Earth. The intensity of the scattered light falls sharply as the Sun dips lower below the horizon, and is almost negligible by the time the Sun's zenith angle reaches 108°, 18° below the horizon. The period after sunset or before sunrise during which the Sun's zenith angle is less than 108° is called *astronomical twilight.*

We can calculate the time at which morning twilight begins or evening twilight ends quite simply. We first find the hour-angles, H, of the Sun at rising or setting by

$$\cos H = -\tan \phi \tan \delta$$

where ϕ is the geographic latitude and δ is the Sun's declination. Then we calculate its hour-angle, H', at the point when its zenith angle is 108° from

$$\cos H' = \frac{\cos 108 - \sin \phi \sin \delta}{\cos \phi \cos \delta}.$$

Then the duration of twilight in hours is simply

$$t = \frac{H' - H}{15} \text{ hours.}$$

During the course of one year the Sun's declination ranges from −23.5 degrees to +23.5 degrees. Latitudes north of +48.5 degrees or south of −48.5 degrees will therefore experience a twilight which lasts all night during the summer. For example, on the latitude 60° N the twilight lasts all night from about April 23rd until August 22nd. When this is so, the value of $\cos H'$ lies outside its allowed range of -1 to $+1$, and your calculator should respond with 'error' if you attempt inverse cos. For example, let us calculate the beginning of morning twilight and the end of evening twilight on September 7th 1979 for an observer at latitude 52° N.

Method	*Example*
1. Calculate the declination of the Sun; its midnight value will do as this is not a very precise calculation (§§ 42 or 43).	$\delta_{\odot} =$ 6.380 389 degrees
2. Find the Sun's hour-angle at setting from $H = \cos^{-1}\{-\tan\phi \tan\delta\}$.	$H =$ 98.228 710 degrees
3. Find the Sun's hour-angle at $z = 108°$ from $H' = \cos^{-1}\left\{\frac{\cos 108 - \sin\phi \sin\delta}{\cos\phi \cos\delta}\right\}$.	$H' =$ 130.404 520 degrees
4. Calculate $t = \frac{H' - H}{15}$ hours.	$t =$ 2.145 054 hours (ST)
5. Multiply t by 0.9973 to convert to an interval of GMT.	$t' =$ 2.139 262 hours (GMT)
6. Having found the time of sunrise and sunset (§ 45), subtract or add the value of t' to find the beginning of morning twilight and the end of evening twilight.	$GMT_r =$ 5.333 hours $GMT_s =$ 18.583 hours morning twilight begins at **03h 12m GMT** evening twilight ends at **20h 43m GMT**

The *Astronomical Ephemeris* gives these times as 03h 17m and 20h 37m, having taken due account of the Sun's changing coordinates throughout the day, refraction and parallax.

47 The equation of time

The apparent motion of the Sun along the plane of the ecliptic is not regular. This is rather surprising at first because we are used to thinking of the Sun as a time-keeper by which we can set our watches. In fact it is really quite a bad time-keeper by modern quartz-crystal watch standards; it can be as much as 16 minutes out compared with a regular clock whose time increases at a uniform rate. The Sun's non-uniform motion is due to two effects:

(*a*) The Earth's orbit is not circular but elliptical. Its speed therefore varies throughout the year being maximum at perihelion and minimum at aphelion. Viewed from the Earth the Sun's speed in its apparent orbit varies from a maximum at perigee to a minimum at apogee.

(*b*) The Earth's axis is tilted at an angle to the perpendicular of the plane of the ecliptic. The angle is the same as the obliquity of the ecliptic, $\varepsilon = 23° 26'$ (section 27). The Earth acts as a huge gyroscope keeping its rotation axis in a fixed direction in space, making the Sun's altitude at noon vary throughout the year from a maximum at midsummer to a minimum at midwinter. This variation in altitude has a small effect on the time of transit of the Sun.

To take account of the Sun's apparent aberrations from perfect time-keeping we imagine a fictitious Sun, called the mean Sun, which moves at a uniform rate along the equator. Noon is defined to be the instant when the mean Sun crosses the meridian, and two successive passages of the mean Sun across it define the length of the day. Time measured by the mean Sun on the Greenwich meridian is called Greenwich mean time. The mean Sun is tied to the real Sun by requiring their positions to be identical at the same instant every year.

The difference between the mean Sun time and the real Sun time is called the equation of time. Hence:

$$E = \text{MST} - \text{RST}$$

where E is the value of the equation of time, MST is the mean Sun time and RST is the real Sun time. It is plotted in Figure 20.

We can calculate the equation of time on any day quite easily by first finding the Sun's right ascension at noon and then, remembering that the right ascension is the sidereal time at transit, converting it to GMT. The result is the GMT at which the real Sun transits; by subtracting this from 12h 00m, the GMT at which the mean Sun transits, we have the value of the equation of time.

Figure 20. Equation of time.

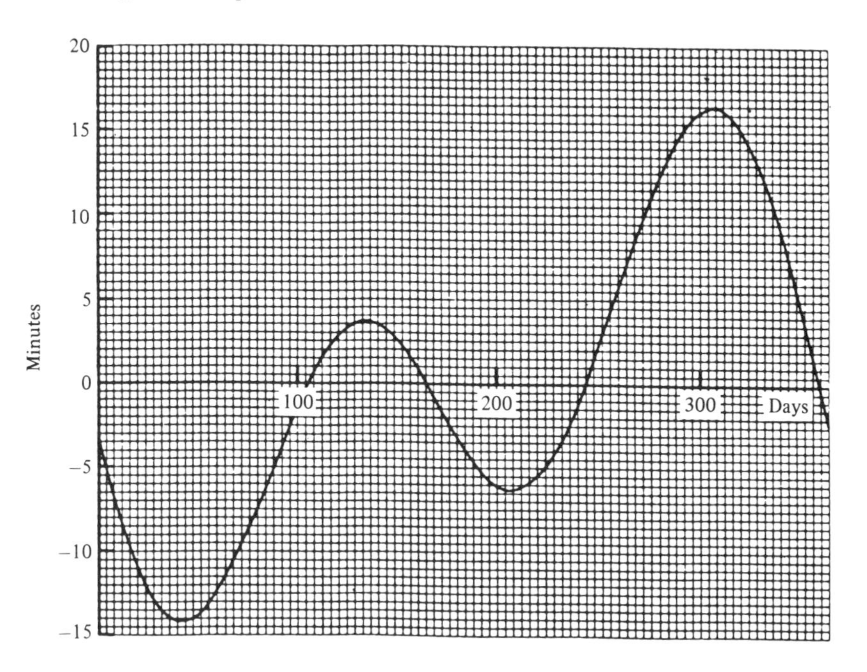

For example, what was the value of the equation of time on July 27th 1980? (Remember that noon is July 27.5).

Method	*Example*
1. Calculate the right ascension of the Sun (§ 42) in decimal hours.	D = 209.5 days α = 8.461 737 hours
2. Taking this as the GST, convert it to GMT (§ 13).	GMT = 12.106 709 hours
3. Subtract this from 12.0 and convert to hours, minutes and seconds (§ 8). This is the value of the equation of time.	E = **−06m 24s**

If you have a sundial, you will need the equation of time to convert the sundial's reading into mean time.

48 Solar elongations

The solar elongation of a planet (or other celestial object) is the angle between the lines of sight from the Earth to the Sun and from the Earth to the planet. It is quite often necessary to find the value of this angle, as it tells us how close to the Sun we should look to see the planet and hence whether it will be visible. The formula for the solar elongation, ε, is:

$$\varepsilon = \cos^{-1}\{\sin\delta_p \sin\delta_\odot + \cos(\alpha_p - \alpha_\odot)\cos\delta_\odot \cos\delta_p\} \text{ degrees}$$

where $\alpha_\odot$ and $\delta_\odot$ are the right ascension and declination of the Sun, and α_p and δ_p are the right ascension and declination of the planet.

On July 27th 1980, the equatorial coordinates of the planet Mercury were found to be $\alpha_p = 07\text{h}\,08\text{m}\,18\text{s}$ and $\delta_p = 19°\,21'\,50''$. What was the solar elongation?

Method	*Example*
1. Calculate $\alpha_\odot$ and $\delta_\odot$ (§§ 42 or 43), leaving the answer in decimal degrees.	$\alpha_\odot = 126.433\,333$ degrees $\delta_\odot = 19.231\,389$ degrees
2. Convert α_p and δ_p into decimal form (§§ 7 and 21).	$\delta_p = 19.363\,889$ degrees $\alpha_p = 7.138\,333$ hours
3. Convert α_p to degrees by multiplying by 15 (§ 22).	$\alpha_p = 107.074\,995$ degrees
4. Find $\varepsilon = \cos^{-1}\{\sin\delta_p \sin\delta_\odot + \cos(\alpha_p - \alpha_\odot)\cos\delta_p \cos\delta_\odot\}$.	$\varepsilon =$ **18.26 degrees**

The planets, comets and binary stars

An observer looking up at the night sky from the surface of the Earth sees an unchanging pattern of stars revolving slowly about the pole as the Earth spins on its axis. So great are the distances to the stars that the changing position of the Earth as it travels along its orbit around the Sun causes hardly any movement in the pattern, even in the course of six months. There are a few objects, however, which do appear to move a great deal with respect to this fixed background of stars. The objects are members of our Solar System, the planets, the asteroids and the comets. Nine major planets have been identified so far which, in order of increasing distance from the Sun, are Mercury, Venus, Earth, Mars, Jupiter, Saturn, Uranus, Neptune and Pluto. These, together with other members of the Solar System, are all bound by the gravitational field of the Sun so that instead of moving off into space in different directions they are constrained to follow elliptical orbits about it. Their apparent motions in the sky are complicated because they are relatively close to us so that the position of the Earth in its own orbit needs to be taken into account. The next few sections contain methods for calculating the positions, angular sizes, distances, phases and brightnesses of the major planets. There are also sections describing how to calculate the orbit of a comet and the orbit of a binary star.

49 The planetary orbits

Each planet in our Solar System describes an elliptical orbit about the Sun with the Sun at a focus of the ellipse (Figure 21). We discovered how to calculate the Sun–Earth orbit in sections 40 to 43. This was a particularly simple case since the plane of the orbit defined the plane of the ecliptic; the ecliptic latitude was therefore always zero and the fundamental direction, the first point of Aries, was in the orbital plane. The other planets, however, do not move in the plane of the ecliptic but describe orbits inclined at small angles to it. Figure 21 shows the situation.

The Sun, S, is at the centre of the diagram and you are to imagine that you are looking at the path of a planet around the Sun from a great distance. The orbit of the planet is the small shaded ellipse N_1AP. The perihelion is marked A and the planet's present position is marked P. That part of the orbit which lies above the ecliptic is shown with solid lines, while

Figure 21. Defining the orbit of a planet.

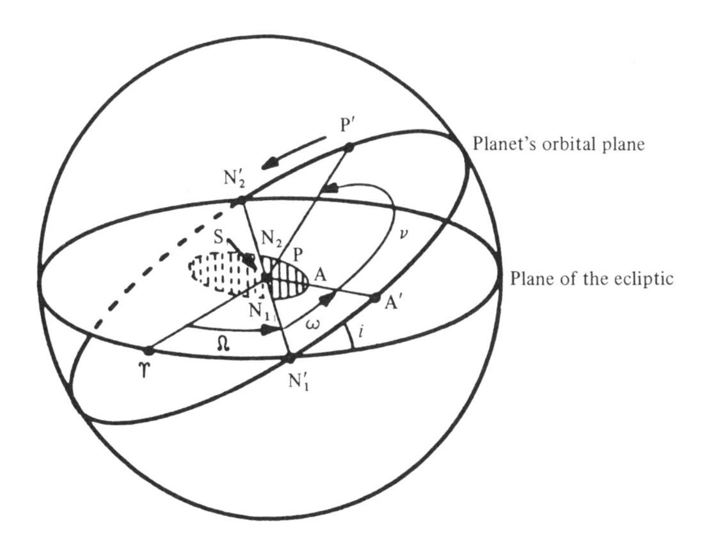

that lying below it is shown with dashed lines. The large sphere is centred on the Sun and the plane of the planet's orbit is projected to cut the sphere along the circle $N_1'A'P'N_2'$. A' is the projection of A onto the sphere, P' the projection of P and so forth. Also shown in the diagram is the plane of the ecliptic ♈$N_1'N_2'$, which contains the direction of the first point of Aries, ♈.

The planet moves along its orbit in the direction of the arrow. The point N_1 where it rises out of the plane of the ecliptic is called the *ascending node*. N_2, the point where it descends below the plane of the ecliptic, is called the *descending node*. Angles in the orbital plane are referred to the ascending node while longitudes are reckoned from the direction ♈ which is not in the orbital plane. Thus the perihelion is at an angle ω to the node (the 'argument' of the perihelion) and the present position of the planet is at an angle $\omega + \nu$. The corresponding longitudes are $\omega + \Omega$ and $\omega + \nu + \Omega$, where Ω is the longitude of the ascending node. Note that longitudes are the sum of two angles in different planes.

50 Calculating the coordinates of a planet

Our calculation will proceed in three steps. The first is to calculate the position of the planet in its own orbital plane exactly as we did for the Sun–Earth orbit in section 42. In the second step we will project the planet's calculated position onto the plane of the ecliptic and hence find its ecliptic longitude and latitude referred to the Sun (heliocentric coordinates). The third step will involve transforming from the Sun to the Earth to find the ecliptic coordinates referred to the Earth, from which we can find the right ascension and declination by the method given in section 27.

As before, we choose our starting point, the epoch, as 1980.0. Having calculated the number of days, D, since the epoch, we find the mean anomaly, M, by the formula

$$M = \frac{360}{365.2422} \times \frac{D}{T_p} + \varepsilon - \varpi \text{ degrees}$$

where T_p is the orbital period of the planet in tropical years, ε is the mean longitude of the planet at the epoch, and ϖ is the longitude of the perihelion. These constants are listed for the planets in our Solar System in Table 7. The mean anomaly refers to the motion of a fictitious planet, P_1, moving in a circle at constant speed with the same orbital period as the real planet (see Figure 22). We really want to know the value of the true anomaly, ν, which is the angle the real planet actually makes with the line joining the Sun to the perihelion. We can find ν from M using the equation of the centre:

$$\nu = M + \frac{360}{\pi} e \sin M \text{ degrees}$$

where e is the eccentricity of the orbit (Table 7) and $\pi =$ 3.141 592 7. Once again, this formula is an approximation which is quite good enough for most purposes; if you wish to make very precise calculations you can find the value of ν by solving Kepler's equation using the method of section 43.

Figure 22. Mean and true anomalies.

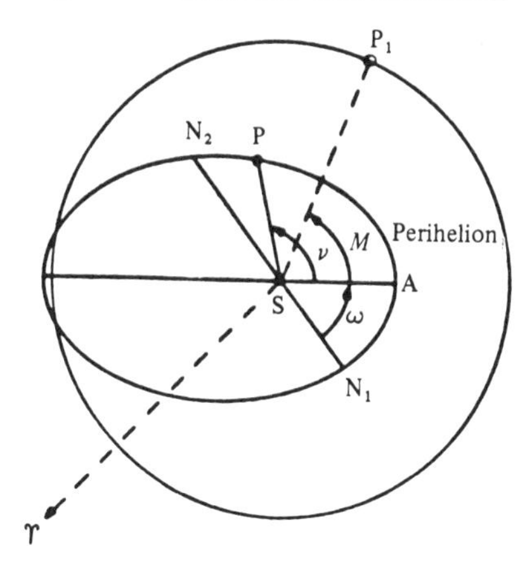

Table 7. Elements of the planetary orbits, epoch 1980.0

	Period, T_p (tropical years)	Longitude at epoch, ε (degrees)	Longitude of perihelion, ϖ (degrees)	Eccentricity of orbit, e	Semi-major axis of orbit referred to Earth's orbit, a (AU)	Inclination of orbit, i (degrees)	Longitude of ascending node, Ω (degrees)	Angular size at 1 AU, θ_0 (arcsec)	Brightness factor, A (AU)^2
Mercury	0.240 85	231.297 3	77.144 212 8	0.205 630 6	0.387 098 6	7.004 357 9	48.094 173 3	6.74	1.918×10^{-6}
Venus	0.615 21	355.733 52	131.289 579 2	0.006 782 6	0.723 331 6	3.394 435	76.499 752 4	16.92	1.721×10^{-5}
Earth	1.000 04	98.833 540	102.596 403	0.016 718	1.000 000	—	—	—	—
Mars	1.880 89	126.307 83	335.690 816 6	0.093 386 5	1.523 688 3	1.849 801 1	49.403 200 1	9.36	4.539×10^{-6}
Jupiter	11.862 24	146.966 365	14.009 549 3	0.048 465 8	5.202 561	1.304 181 9	100.252 017 5	196.74	1.994×10^{-4}
Saturn	29.457 71	165.322 242	92.665 397 4	0.055 615 5	9.554 747	2.489 374 1	113.488 834 1	165.60	1.740×10^{-4}
Uranus	84.012 47	228.070 855 1	172.736 328 8	0.046 323 2	19.218 14	0.772 989 5	73.876 864 2	65.80	7.768×10^{-5}
Neptune	164.795 58	260.357 899 8	47.867 214 8	0.009 002 1	30.109 57	1.771 601 7	131.560 649 4	62.20	7.597×10^{-5}
Pluto*	250.9	209.439	222.972	0.253 87	39.784 59	17.137	109.941	8.20	4.073×10^{-6}

1 AU $= 149.6\times10^6$ km.

* The elements for Pluto are the *osculating elements* for epoch 1980 January 2.0.

The next step is to calculate the heliocentric longitude, l; and this is simply given by

$$l = \nu + \varpi$$

or

$$l = \left(\frac{360}{365.2422} \times \frac{D}{T_p}\right) + \frac{360}{\pi} e \sin\left(\frac{360}{365.2422} \times \frac{D}{T_p} + \varepsilon - \varpi\right) + \varepsilon \text{ degrees.}$$

We also need the length of the radius vector, r, calculated from

$$r = \frac{a(1 - e^2)}{1 + e \cos \nu}$$

where a is the semi-major axis of the orbit (Table 7).

The above calculations which you have made for the planet have to be repeated for the Earth as well. We shall denote the values derived for the planet by small letters and use capital letters for the Earth's values. Thus, we arrive at the figures for l and r for the planet and L and R for the Earth. In addition, we need the heliocentric latitude of the planet:

$$\psi = \sin^{-1}\{\sin(l - \Omega) \sin i\},$$

where i is the inclination of the orbit and Ω the longitude of the ascending node (Table 7). The heliocentric latitude of the Earth is, of course, zero.

Now we need to project our calculations for the planet onto the plane of the ecliptic to find the projected heliocentric longitude, l', and the projected radius vector, r'. These are given by the formulae

$$l' = \tan^{-1}\{\tan(l - \Omega) \cos i\} + \Omega,$$

$$r' = r \cos \psi.$$

The final step in the process is to refer the calculations to the Earth to find the geocentric ecliptic latitude, β, and longitude, λ, of the planet. Figure 23a describes the situation for a planet whose orbit lies outside that of the Earth, and Figure 23b is for an inner planet (the inner planets are Mercury and Venus).

Figure 23. Ecliptic geometry: (*a*) outer planet, (*b*) inner planet.

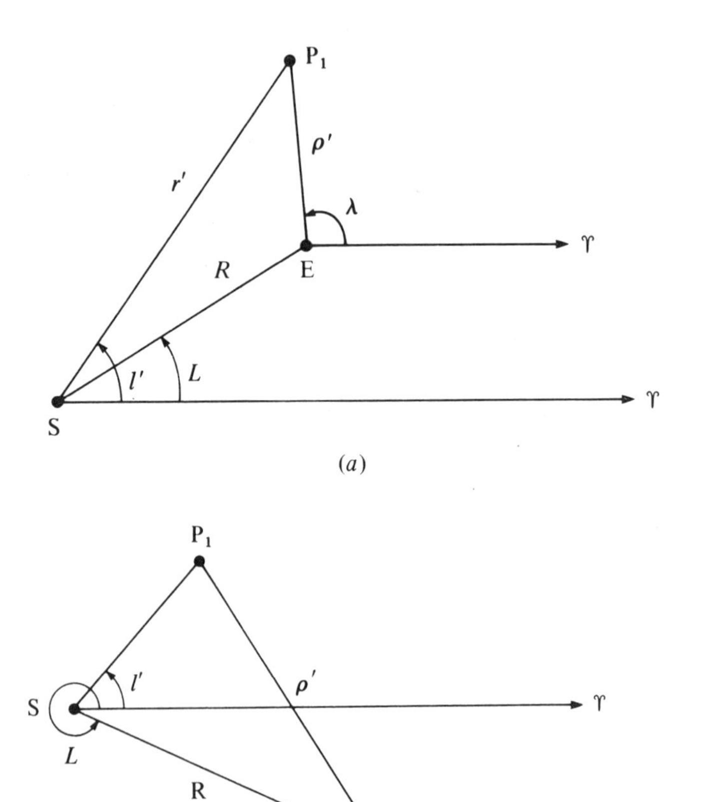

The plane of the paper represents the plane of the ecliptic. S is the Sun, E is the earth and P_1 is the position of the planet projected onto the ecliptic. The first point of Aries is taken to be at a distance from the Solar System so large that the directions E♈ and S♈ are parallel. Then by application of a little simple geometry we have for the outer planets

$$\lambda = \tan^{-1}\left\{\frac{R \sin(l' - L)}{r' - R\cos(l' - L)}\right\} + l' \text{ degrees,}$$

and for the inner planets

$$\lambda = 180 + L + \tan^{-1}\left\{\frac{r' \sin(L - l')}{R - r'\cos(L - l')}\right\} \text{ degrees.}$$

Figure 24 gives the diagram for calculating the latitude. Again, using simple geometry we find

$$\beta = \tan^{-1}\left\{\frac{r' \tan\psi \sin(\lambda - l')}{R \sin(l' - L)}\right\} \text{ degrees,}$$

true for inner and outer planets alike.

Let us consolidate these rather lengthy calculations with two examples, one for an inner planet, Mercury, and the other for an outer planet, Jupiter. For each planet we shall calculate its right ascension and declination on November 22nd 1980.

Figure 24. Projecting onto the ecliptic.

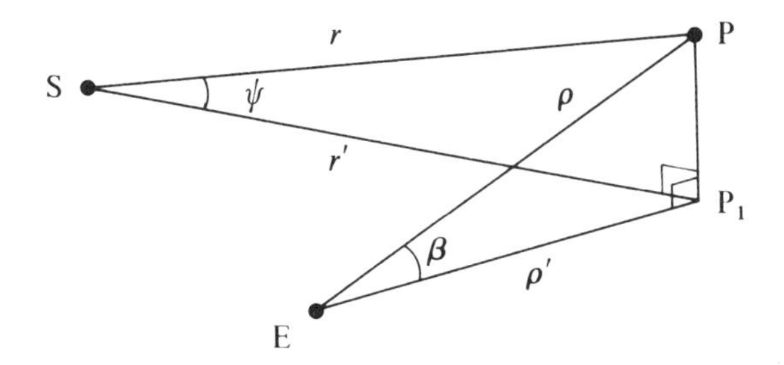

For Mercury (inner planet):

Method	*Example*
	November 22nd
1. Find the number of days since January 0.0 (§ 3). Add 365 for each year since 1980, plus one extra day for each leap year (see Table 2). The total is D.	$= 305+22$ days $+ \quad 0$ $D = 327$ days
For the planet	
2. Calculate $N_p = \frac{360}{365.2422} \times \frac{D}{T_p}$; subtract multiples of 360 to bring the result into the range 0–360.	$N_p = 1338.205\,014$ degrees -360×3 $= 258.205\,014$ degrees
3. Find $M_p = N_p + \varepsilon - \varpi$.	$M_p = 412.358\,101$ degrees
4. Calculate $l = N_p + \frac{360}{\pi} e \sin(M_p) + \varepsilon$ ($\pi = 3.141\,592\,7$). If the result is more than 360, subtract 360. If the result is negative, add 360.	$l = 148.160\,937$ degrees
5. Find $\nu_p = l - \varpi$.	$\nu_p = 71.016\,724$ degrees
6. Calculate $r = \frac{a(1-e^2)}{1+e\cos\nu_p}$ (ratio to Earth's orbit).	$r = 0.347\,487$ AU
Now do the calculations for the Earth	
7. Find $N_E = \frac{360}{365.2422} \times \frac{D}{T_E}$. Subtract multiples of 360 to bring the result into the range 0–360.	$N_E = 322.293\,786$ degrees
8. Calculate $M_E = N_E + \varepsilon - \varpi$.	$M_E = 318.530\,923$ degrees
9. Find $L = N_E + \frac{360}{\pi} e \sin(M_E) + \varepsilon$ ($\pi = 3.141\,592\,7$). If the result is more than 360, subtract 360. If the result is negative, add 360.	$L = 59.858\,692$ degrees
10. Calculate $\nu_E = L - \varpi$.	$\nu_E = -42.737\,712$ degrees
11. Find $R = \frac{(1-e^2)}{1+e\cos\nu_E}$ (ratio to Earth's orbit).	$R = 0.987\,594$ AU
12. Calculate $\psi = \sin^{-1}\{\sin(l-\Omega)\sin i\}$.	$\psi = 6.895\,996$ degrees
13. Find $y = \sin(l-\Omega)\cos i$.	$y = 0.977\,257$
14. Find $x = \cos(l-\Omega)$.	$x = -0.174\,796$

Method (continued)	*Example*
15. Calculate $\tan^{-1}\left\{\frac{y}{x}\right\}$.	$\tan^{-1}\left\{\frac{y}{x}\right\} = -79.859\,090$ degrees
It is necessary to remove the ambiguity on taking inverse tan. Look at the signs of x and y and compare with Figure 10. Add or subtract 180 or 360 until $\tan^{-1}\left\{\frac{y}{x}\right\}$ lies in the correct quadrant (unless it is already in the correct quadrant).	y positive x negative $+$ 180.0 $=$ 100.140 911 degrees
16. Add ☊ to get l' (check: l' should be nearly equal to l).	$l' =$ 148.235 084 degrees
17. Find $r' = r\cos\psi$.	$r' =$ 0.344 973 degrees
18. Find $A = \tan^{-1}\left\{\frac{r'\sin(L-l')}{R-r'\cos(L-l')}\right\}$.	$L-l' =$ −88.376 392 degrees $A =$ −19.425 531 degrees
19. Calculate $\lambda = 180 + L + A$. If the answer is negative, add 360. If the answer is more than 360, subtract 360. This is the planet's geocentric longitude.	$\lambda =$ 220.433 161 degrees
20. Find the geocentric latitude, $\beta = \tan^{-1}\left\{\frac{r'\tan\psi\sin(\lambda-l')}{R\sin(l'-L)}\right\}$.	$\beta =$ 2.304 301 degrees
21. Finally, calculate the right ascension and declination using the method given in § 27.	$\alpha =$ **14h 35m 02s** $\delta =$ **−12° 45′ 46″**

The *Astronomical Ephemeris* gives these coordinates as $\alpha = 14$h 35m 30s and $\delta = -12° 50' 29''$. Even on the planet Mercury with its highly elliptical orbit this method gives quite a good result. We should generally expect an error in α of a few minutes at most and in δ of a quarter of a degree, but the errors may be more for Mercury, for which $e = 0.2$. The inaccuracies arise because we have used only the first term in the equation of the centre and because of the slight perturbations to the orbits from other planets in the Solar System (see section 52). We could reduce the error from the first cause by using the longer method of section 43; see Figure 25 (section 52) for a plot of the error incurred by the shorter method.

For Jupiter (outer planet):

Method	*Example*
1. We proceed exactly as in the previous example calculating l, ν_p, r, ψ, l' and r' for Jupiter, and L, ν_E and R for the Earth.	$l =$ 176.025 048 degrees $\nu_p =$ 162.015 499 degrees $r =$ 5.441 166 *AU* $\psi =$ 1.264 176 degrees $l' =$ 176.021 511 degrees $r' =$ 5.439 842 AU $L =$ 59.858 692 degrees $\nu_E =$ −42.737 712 degrees $R =$ 0.987 594 AU
2. Now calculate $\lambda = \tan^{-1}\left\{\frac{R \sin(l'-L)}{r' - R\cos(l'-L)}\right\} + l'$. If the answer is more than 360, subtract 360. If the answer is negative, add 360.	$(l'-L) =$ 116.162 819 degrees $\lambda =$ 184.601 058 degrees
3. Find $\beta = \tan^{-1}\left\{\frac{r' \tan\psi \sin(\lambda - l')}{R \sin(l'-L)}\right\}$.	$\beta =$ 1.157 413 degrees
4. Convert to right ascension and declination using the method of § 27.	$\alpha =$ **12h 18m 44s** $\delta =$ **−0° 45′ 59″**

The *Astronomical Ephemeris* gives the coordinates of Jupiter for this day as $\alpha = 12\text{h}\ 18\text{m}\ 51\text{s}$ and $\delta = -0°\ 46'\ 40''$. Again the error due to our approximation in counting only the first term of the equation of the centre could be reduced by solving Kepler's equation using the method in section 43. The most important perturbations due to other planets in the Solar System can also be allowed for; see section 52.

51 Finding the approximate positions of the planets

The method of finding the equatorial coordinates of the planets given in the previous section is quite accurate but involves lengthy calculations. An amateur astronomer often only wants to know the approximate position of a planet so that he knows where to look for it in the sky, and does not want to have to spend 20 minutes beforehand submerged in a sea of figures obtaining the information. In that case it is usually sufficient to assume that the planets describe circular orbits about the Sun which lie in the plane of the ecliptic. This leads to considerable simplifications in the calculations.

The heliocentric longitude, l, does not have to be corrected by the equation of the centre so that we may write

$$l = \frac{360}{365.2422} \times \frac{D}{T_p} + \varepsilon \text{ degrees.}$$

We repeat this calculation for the Earth as before (giving L). Since the orbits are assumed to be circular with the Sun at the centre the radius vector is constant. Hence

$$r = a.$$

The heliocentric (and therefore the geocentric) latitude of the planet is zero since we have assumed that the orbit lies in the ecliptic plane. Our final calculation is therefore

$$\lambda = \tan^{-1}\left\{\frac{\sin(l-L)}{a-\cos(l-L)}\right\} + l$$

for the outer planets and

$$\lambda = 180 + L + \tan^{-1}\left\{\frac{a\sin(L-l)}{1-a\cos(L-l)}\right\}$$

for the inner planets, since $R = 1$ (the Earth's orbital radius is taken to be unity). This is the geocentric ecliptic longitude of the planet from which the right ascension and declination can be found using the formulae of section 27 (remember $\beta = 0$):

$$\alpha = \tan^{-1}\{\tan\lambda\cos\varepsilon\},$$

$$\delta = \sin^{-1}\{\sin\varepsilon\sin\lambda\},$$

where ε here is the obliquity of the ecliptic (see section 27). In some cases it may even be possible to ignore the fact that the plane of the ecliptic is inclined at an angle to the plane of the equator and to write

$$\alpha = \lambda.$$

For our example, we will calculate again the coordinates of Jupiter on November 22nd 1980 using this approximate method.

Method	*Example*
	November 22nd
1. Find the number of days since January 0.0 (§ 3). Add 365 for each year since 1980 plus one day for each leap year (see Table 2). The total is D.	$= 305 + 22$ days $+ \quad 0$ $D = 327$ days
2. Calculate $l = \dfrac{360}{365.2422} \times \dfrac{D}{T_p} + \varepsilon$ and subtract multiples of 360 to bring the result into the range 0–360.	$l = 174.14$ degrees
3. Repeat step 2 for the Earth to find L.	$L = \ 61.13$ degrees
4. Calculate $\lambda = \tan^{-1}\left\{\dfrac{\sin(l-L)}{a-\cos(l-L)}\right\} + l.$	$\lambda = 183.48$ degrees
5. Convert to right ascension and declination (§ 27) remembering that $\beta = 0$.	$\alpha =$ **12h 13m** $\delta =$ **−1° 23′**

The result is within 6 minutes in α and half a degree in δ of the true position. We must expect, however, that the errors may be much larger than this in some cases, especially where the orbit has a high value of e or i.

52 Perturbations in a planet's orbit

Throughout the calculations to find the coordinates of a planet (section 50), we assumed that its motion was controlled entirely by the gravitational field of the Sun so that the influences of other members of the Solar System were negligible. This is true to quite a high accuracy, but for more precision we need to take account of these *perturbations*, especially for the orbits of Jupiter and Saturn where the effects can be as large as one degree in longitude. The usual method of doing so is to apply a series of correction terms to the quantities calculated in section 50. We have to make similar adjustments for the Moon in section 61. Here, we shall consider only the most important terms in the orbits of Jupiter and Saturn where the corrections amount to more than about 0.04 degrees in longitude.

We must first calculate the time, T, in Julian centuries since the epoch 1900 January 0.5. This is given by

$$T = \frac{\mathrm{JD} - 2\,415\,020.0}{36\,525},$$

where JD is the Julian date (section 4). Then we calculate the quantities:

$$A = \frac{T}{5} + 0.1,$$

$$P = 237°.475\,55 + 3034°.9061\,T,$$

$$Q = 265°.916\,50 + 1222°.1139\,T,$$

$$V = 5Q - 2P,$$

and

$$B = Q - P.$$

The principal terms for Jupiter and Saturn are then:

Jupiter:

$$\Delta l = (0.3314 - 0.0103A)\sin V - 0.0644A\cos V.$$

Saturn:

$$\begin{aligned}\Delta l = {} & (0.1609A - 0.0105)\cos V \\ & + (0.0182A - 0.8142)\sin V \\ & - 0.1488\sin B - 0.0408\sin 2B \\ & + 0.0856\sin B\cos Q \\ & + 0.0813\cos B\sin Q.\end{aligned}$$

The value of Δl must be added to the mean longitude l before proceeding with the calculation of section 50.

Let us now recalculate the position of Jupiter on November 22nd 1980, solving Kepler's equation properly (section 43) and allowing for these principal terms of perturbation.

Method	*Example*
1. Calculate the Julian date (§ 4).	JD = 2 444 565.50 days
2. Find $T = \dfrac{\text{JD} - 2\,415\,020.0}{36\,525}$.	$T =$ 0.808 912 centuries
3. Find $A = \dfrac{T}{5} + 0.1$.	$A =$ 0.261 782 centuries
4. Find $P = 237°.475\,55 + 3034°.9061T$.	$P =$ 2 692.447 513 degrees
5. Find $Q = 265°.916\,50 + 1222°.1139T$.	$Q =$ 1 254.499 099 degrees
6. Find $V = 5Q - 2P$.	$V =$ 887.600 469 degrees
7. Calculate $\Delta l = (0.3314 - 0.0103A) \sin V - 0.0644A \cos V$.	$\Delta l =$ 0.087 047 degrees
8. Now proceed as in the example of § 50 to find M_p.	$M_p =$ 160.127 626 degrees
9. Use the method of § 43 to find ν_p.	$E_p =$ 2.810 509 radians
	$\nu_p =$ 161.913 246 degrees
10. Find $l = \nu_p + \varpi$.	$l =$ 175.922 796 degrees
11. Add Δl to get a better estimate of l.	+ 0.087 047 degrees
	$\therefore l_p =$ 176.009 843 degrees
12. Calculate L, ν_E and R for the Earth also using the method of § 43.	$E_E =$ 5.548 202 radians
	$\nu_E =$ −42.757 812 degrees
	$L =$ 59.838 591 degrees
	$R =$ 0.987 598 AU
	$r =$ 5.441 013 AU
13. Now proceed with the calculations of § 50 to find α and δ.	$\psi =$ 1.264 091 degrees
	$l' =$ 176.006 303 degrees
	$r' =$ 5.439 689 AU
	$(l' - L) =$ 116.167 712 degrees
	$\lambda =$ 184.588 480 degrees
	$\beta =$ 1.157 401 degrees
	$\alpha =$ **12h 18m 41s**
	$\delta =$ **−0° 45′ 41″**

We see that, despite our efforts, the answers seem to be further from their true values. This illustrates an important point: although the method given in section 50 appeared to give a better answer in this case, it has a larger overall error associated with it and would generally be less accurate. The error incurred in considering only the first term of the equation of the centre is plotted as a function of the mean anomaly, M, in Figure 25 for two values of the eccentricity, e. In this example, the error was 0.102 degrees which, by chance, nearly cancelled with other errors to give us a close result.

Figure 25. The error, Δ, incurred by taking $\nu = M + (360/\pi)e \sin M$ as an approximation to elliptical motion. The true anomaly should be $\nu + \Delta$.

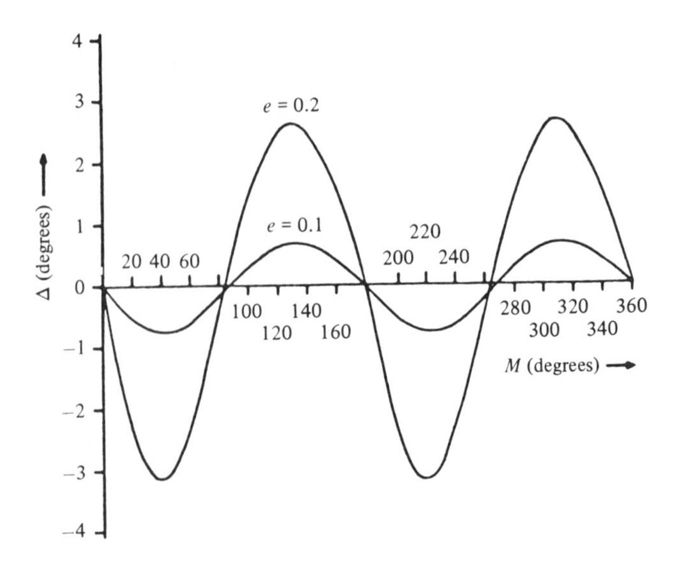

53 The distance, light-travel time and angular size of a planet
During the course of our calculations in section 50 to determine the position of a planet we found the distances r and R of the planet and the Earth, respectively, from the Sun. We can quite easily use these values together with the heliocentric longitudes l and L to calculate the planet's distance from the Earth, ρ. The situation is drawn in Figure 26. We assume, with negligible error, that the planet, P, lies in the plane of the ecliptic. Then by applying the cosine rule to the triangle SEP we have

$$\rho^2 = R^2 + r^2 - 2Rr\cos(l - L),$$

from which ρ may be found. It is usual to express r and R in astronomical units (AU) where 1 AU is the semi-major axis of the Earth's orbit. ρ is then the distance of the planet from the Earth measured in AU.

Having calculated ρ, it is then an easy matter to find the light-travel time, τ, the time taken for the light to reach us

Figure 26. Finding the distance of a planet.

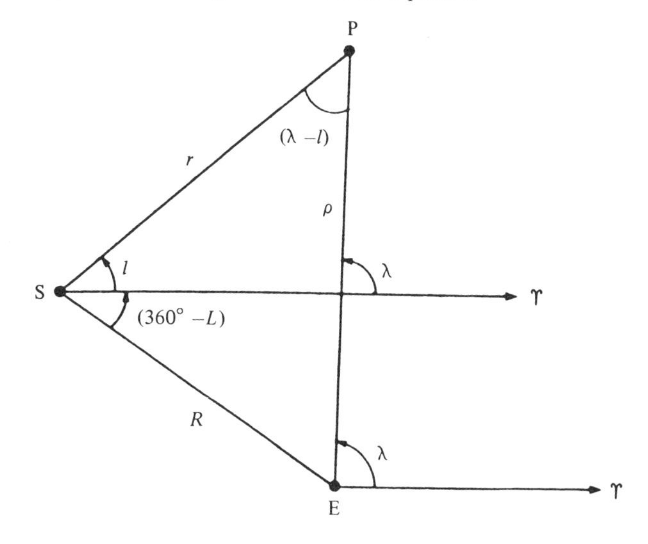

from the planet. When we view a planet now, we see it in the position it occupied τ hours ago, given by

$$\tau = 0.1386\rho \text{ hours},$$

where ρ is expressed in AU.

We can also find the apparent angular diameter, θ, of the planet given by

$$\theta = \frac{\theta_0}{\rho},$$

where ρ is again expressed in AU and θ_0 is the angular diameter of the planet when it is at 1 AU from the Earth. Values of θ_0 are given in Table 7.

We shall calculate the distance, the light-travel time, and the apparent angular diameter of Jupiter on November 22nd 1980.

Method	*Example*
1. Find r, R, l and L as in § 50.	$r =$ 5.441 166 AU
	$R =$ 0.987 594 AU
	$l =$ 176.025 048 degrees
	$L =$ 59.858 692 degrees
2. Calculate $\rho^2 = R^2 + r^2 - 2Rr\cos(l - L)$.	$\rho^2 =$ 35.320 973 AU^2
3. Take the square root to find ρ.	$\rho =$ **5.94 AU**
4. Multiply by 0.1386 to find τ; convert	$\tau =$ 0.823 720 hours
to minutes and seconds form (§ 8).	$\tau =$ **49m 25s**
5. Find θ from $\theta = \dfrac{\theta_0}{\rho}$.	$\theta =$ **33 arcsec**

The *Astronomical Ephemeris* quotes $\rho = 5.94$ AU and $\theta =$ 31 arcsec for Jupiter this day.

54 The phases of the planets

At any point in the orbit of a planet, the hemisphere which faces towards the Sun is brightly illuminated while the other half of the planet's surface is dark. The fraction of the surface which we can see from the Earth, however, is that part lying on the hemisphere facing the Earth which usually overlaps both the bright and the dark sides. We are presented therefore with a view of the planet's disc which is not uniformly illuminated but which contains a bright segment, the rest of the disc being dark and usually invisible. As the relative positions of the Earth, the planet and the Sun vary, so the area of the visible disc that is illuminated changes. The phase is defined to be the fraction of the visible disc that is illuminated.

In Figure 26, the angle $(\lambda - l)$ at P is the solar elongation of the Earth as measured at the planet. We represent this angle by d. Thus

$$d = \lambda - l.$$

The phase, F, is related to d by the formula

$$F = \tfrac{1}{2}(1 + \cos d).$$

F always lies in the range 0 to 1. When $F = 0$ the whole of the dark side of the planet is towards the Earth. This can only happen for the inner planets Mercury and Venus. When $F = 1$ the whole of the bright side faces the Earth.

We shall find the phases of Mercury and Jupiter on November 22nd 1980 as our example.

Method	*Example*
1. Calculate $d = \lambda - l$ using the method outlined in § 50 to find λ and l.	Mercury: $d_1 = 72.272\,224$ Jupiter: $d_2 = 8.576\,010$
2. Find $F = \frac{1}{2}(1 + \cos d)$.	Mercury: $F_1 =$ **0.65** Jupiter: $F_2 =$ **0.99**

We see that practically the whole of the Jovian disc was illuminated while little more than half of that of Mercury was bright.

55 The position-angle of the bright limb

Figure 27 shows the appearance of a planet whose phase is about $F = 0.7$. The dashed outline is of that part of the disc which is invisible, and the line NS is the projection of the Earth's axis onto the disc. The terminator, the line dividing night from day, is the curve AB. Position-angles are measured anticlockwise from the North. Thus points A and B are at position-angles θ_1 and θ_2. The point C, halfway between A and B on the circumference of the disc, is the midpoint of the bright side and it defines the position-angle, χ, of the bright limb. Hence

$$\chi = \tfrac{1}{2}(\theta_1 + \theta_2).$$

We can easily calculate χ provided we know the equatorial coordinates of the planet (α, δ) and of the Sun $(\alpha_\odot, \delta_\odot)$. Then

$$\chi = \tan^{-1}\left\{\frac{\cos\delta_\odot \sin(\alpha_\odot - \alpha)}{\cos\delta \sin\delta_\odot - \sin\delta \cos\delta_\odot \cos(\alpha_\odot - \alpha)}\right\}.$$

Figure 27. The position-angle of the bright limb.

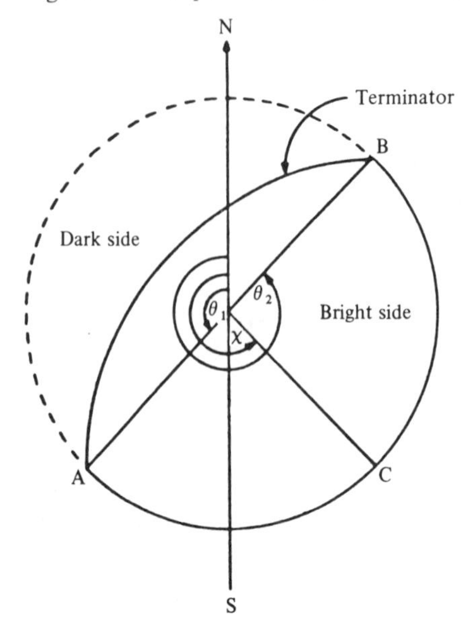

For example, what was the position-angle of the bright limb of Mercury on November 22nd 1980? The Sun's coordinates were $\alpha_\odot = 15\text{h}\ 50\text{m}\ 37\text{s}$, $\delta_\odot = -20°\ 07'\ 04''$.

Method	*Example*
1. Find the right ascension and declination of the planet (§ 50).	$\alpha =$ 14h 35m 02s $\delta = -12°\ 45'\ 46''$
2. Convert both these and the coordinates of the Sun to decimal form (§§ 7 and 21). The coordinates of the Sun may be found from § 42.	$\alpha =$ 14.583 889 hours $\delta = -12.762\ 778$ degrees $\alpha_\odot =$ 15.843 611 hours $\delta_\odot = -20.117\ 778$ degrees
3. Find $\Delta\alpha = \alpha_\odot - \alpha$. Convert to degrees by multiplying by 15 (§ 22).	$\Delta\alpha =$ 18.895 830 degrees
4. Find $y = \cos\delta_\odot \sin\Delta\alpha$.	$y =$ 0.303 973
5. Find $x = \cos\delta \sin\delta_\odot - \sin\delta \cos\delta_\odot \cos\Delta\alpha$.	$x = -0.139\ 196$
6. Find $\chi' = \tan^{-1}\left\{\frac{y}{x}\right\}$. We have to remove the ambiguity here from taking inverse tan. To do so, look up Figure 10 and compare the signs of x and y. Add or subtract 360 or 180 to χ' to bring it into the correct quadrant. If it is already in the correct quadrant, $\chi = \chi'$.	$\chi' = -65.395\ 952$ + 180.0 $\chi =$ **114.6 degrees**

The *Astronomical Ephemeris* quotes $\chi = 114°$ for Mercury on this day.

56 The apparent brightness of a planet

Our calculations so far have given us the position, the solar elongation (section 48), the distance from the Earth, the apparent angular diameter, the phase, and the position-angle of the bright limb of a planet. We need only add the apparent brightness to the list to obtain all the important parameters of the planet's visual aspect.

Brightness is usually measured in *magnitudes*, m, on a non-linear scale such that decreasing brightness goes with increasing magnitude. The brightest stars have an apparent magnitude of about 1 while the faintest stars just visible with the unaided eye are of magnitude 6. The ratio in the light power flux between one magnitude and the next is about 2.5. The Sun, very bright at the Earth, has a visual magnitude of −26.74 while the Moon's magnitude at opposition is −12.73. The planets range from about $m = -4$ for Venus at its most brilliant to $m = +14$ for Pluto at its brightest.

The variation in a planet's brightness is caused by several factors. First the Sun's light flux on the planet varies inversely as the square of its distance, r, from the Sun. Then the amount of that light reradiated towards the Earth depends on the phase, F, and a 'brightness factor', A, the latter being a measure of the reflectivity of the planet combined with the area of the planet's disc. The larger the planet's area, the more light it intercepts from the Sun and hence the more it radiates towards the Earth. Finally, the light flux received from the planet varies inversely as the square of the planet's distance, ρ, from the Earth.

We can obtain an approximate value for the apparent magnitude of a planet from the formula

$$m = 5 \log_{10}\left[\frac{r\rho}{A\sqrt{F}}\right] - 26.7,$$

where r and ρ are measured in AU. The values of A are listed in Table 7. As an example, let us calculate the apparent magnitude of Jupiter on November 22nd 1980.

Method	*Example*
1. Find the values of r, ρ and F using the methods given in §§ 50, 53 and 54.	$r =$ 5.441 166 AU $\rho =$ 5.94 AU $F =$ 0.99
2. Calculate $m = 5 \log_{10}\left[\frac{r\rho}{A\sqrt{F}}\right] - 26.7$. r and ρ must be expressed in AU.	$m =$ **−0.64**

The value of m given in the *Astronomical Ephemeris* for Jupiter on November 22nd is $m = -1.4$. In general, our calculations should be correct to within a magnitude or so. No account has been taken of *atmospheric extinction* (see section 39) which can increase the apparent magnitude of a star or planet near the horizon by 2 or 3. None the less, our calculations will provide a fair guide of what to expect.

57 Comets

In earlier sections we discovered how to calculate the orbit of any solid body moving around a central massive object, and we applied the method to the major planets of our Solar System. All we needed to know were the orbital elements of each planet. Likewise, we can calculate the position of a periodic comet given its orbital elements but the method needs to be modified slightly for two reasons:

(i) The longitude of the comet is not usually specified at a particular epoch. Rather, the epoch is given when the comet is at perihelion, the point of its closest approach to the Sun.

(ii) The eccentricity, e, of a comet is usually much more than 0.1 so that the equation of the centre does not apply. Instead, we have to solve Kepler's equation properly.

The orbital elements of some periodic comets are given in Table 8. Note that, as in the case of the planetary elements, we have specified ϖ, the longitude of the perihelion. Sometimes the *argument* of the perihelion is given instead. It has the symbol ω (very confusing) and is related to ϖ by $\varpi = \omega + \Omega$.

We begin, as before, by finding the mean anomaly, M, of the comet given by the formula

$$M = \frac{360}{365.2422} \times \frac{D}{T_{\mathrm{p}}} + \varepsilon - \varpi,$$

where D is the number of days since the epoch, T_{p} is the orbital period in years, ε is the longitude at the epoch, and ϖ is the longitude of the perihelion. In this case, however, the epoch is the moment of perihelion so that $\varepsilon = \varpi$. Further, we probably do not need to specify the date so accurately in terms of days since the epoch. Rather, we can work in decimal years. Hence, M may be found from

$$M = \frac{360Y}{T_{\mathrm{p}}},$$

where Y is the number of years since perihelion.

Table 8. The orbital elements of some periodic comets

Comet name	Perihelion epoch	Perihelion longitude, ϖ (degrees)	Longitude of ascending node, Ω (degrees)	Period, T_p (years)	Semi-major axis of orbit, a (AU)	Eccentricity of orbit, e	Inclination of orbit, i (degrees)
Encke	1974.32	160.1	334.2	3.30	2.209	0.847	12.0
Temple 2	1972.87	310.2	119.3	5.26	3.024	0.549	12.5
Haneda-Campos	1978.77	12.016	131.700	5.37	3.066	0.64152	5.805
Schwassmann-Wachmann 2	1974.70	123.3	126.0	6.51	3.489	0.386	3.7
Borrelly	1974.36	67.8	75.1	6.76	3.576	0.632	30.2
Whipple	1970.77	18.2	188.4	7.47	3.821	0.351	10.2
Oterma	1958.44	150.0	155.1	7.88	3.958	0.144	4.0
Schaumasse	1960.29	138.1	86.2	8.18	4.054	0.705	12.0
Comas Sola	1969.83	102.9	62.8	8.55	4.182	0.577	13.4
Schwassmann-Wachmann 1	1974.12	334.1	319.6	15.03	6.087	0.105	9.7
Neujmin 1	1966.94	334.0	347.2	17.93	6.858	0.775	15.0
Crommelin	1956.82	86.4	250.4	27.89	9.173	0.919	28.9
Olbers	1956.46	150.0	85.4	69.47	16.843	0.930	44.6
Pons-Brooks	1954.39	94.2	255.2	70.98	17.200	0.955	74.2
Halley	1986.112	170.0110	58.1540	76.0081	17.9435	0.9673	162.2384

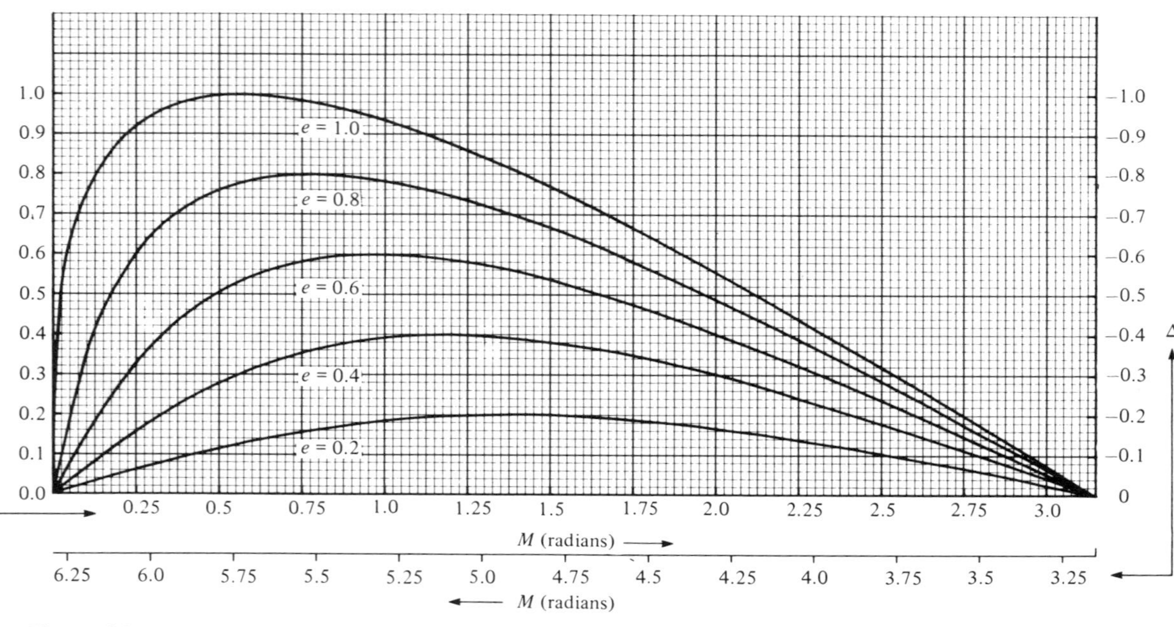

Figure 28. Kepler's graphs. Use the left-hand and upper M scales for values of M between 0 and 3.14, and the right-hand and lower M scales for values of M between 3.14 and 6.28.

Next we have to solve Kepler's equation

$$E - e \sin E = M,$$

where e is the eccentricity and E is the eccentric anomaly. A method of doing this was given in section 43, routine R2, in which the eccentricity was assumed to be less than 0.1 so that the first guess at the solution, $E = M$, was good enough for an accurate solution to be reached after only one or two iterations. Here, the eccentricity is much larger and although the routine would always converge eventually, many iterations might be needed. We can speed up the process if we know an approximate solution to begin with, better than $E = M$. Kepler's graphs, Figure 28, are provided for this purpose. Given any value of e between 0 and 1 and the value of M (expressed in radians), you choose the corresponding value of Δ from the graphs. Then, in place of the first guess $E = E_0 = M$, use $E = E_0 = M + \Delta$ and proceed with the rest of routine R2 as before. You should find that only two or three iterations are needed whatever the values of e and M.

Alternatively, you may like to use the nomogram of Figure 29 to find Δ. Place a ruler across the diagram joining the value of M (in radians) on the right-hand vertical scale with the value of e on the left-hand vertical scale. The point of intersection with the curve gives the magnitude of Δ/e. Multiply this by e to find Δ and give it the sign shown on the right-hand scale. For example, the line joining $M = 5.6$ with $e = 0.46$ cuts the curve at $|\Delta/e| = 0.9$. Thus $|\Delta| = 0.9 \times 0.46 = 0.41$ and its sign is negative, giving

$$\Delta = -0.41.$$

Having found E, we can calculate the true anomaly, ν, from

$$\tan \frac{\nu}{2} = \left[\frac{1+e}{1-e}\right]^{\frac{1}{2}} \tan \frac{E}{2}$$

(all angles in *radians*), and then carry on with the rest of the calculations of section 50. If we find that r' is less than R, we must use the formula at the end which is appropriate for an

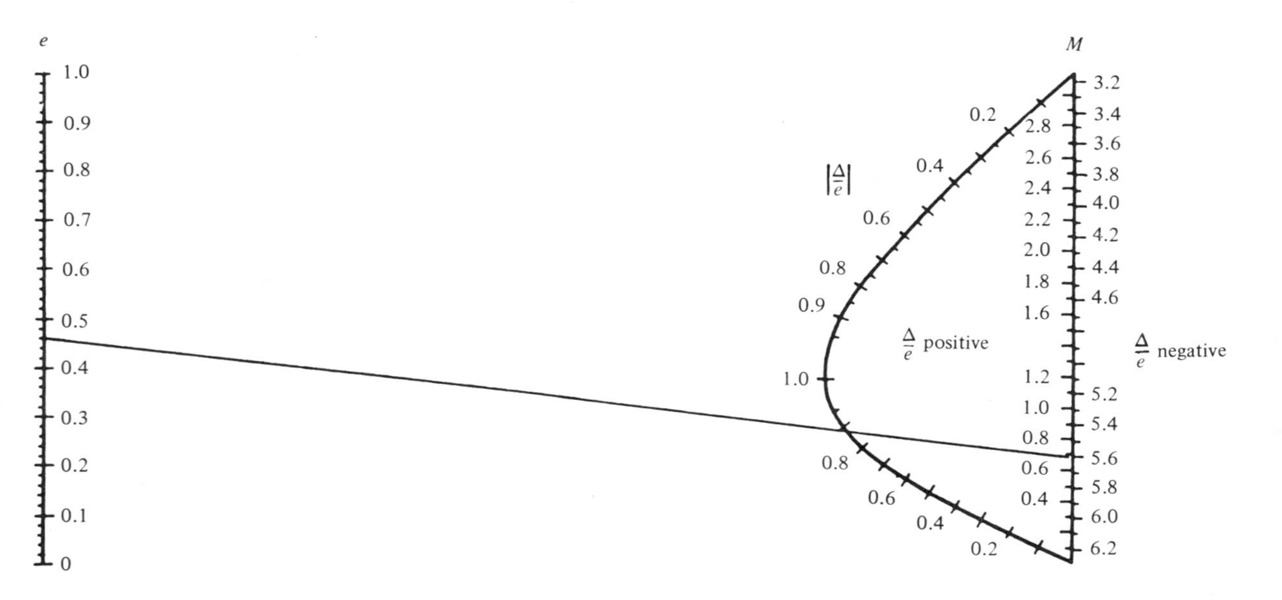

Figure 29. Nomogram to calculate Δ.

inner planet, while if r' is greater than R, we must use the formula for an *outer* planet. In these calculations, remember that the epoch for the comet and the epoch for the Earth are usually different.

The method of finding the position of a comet is best clarified by an example. Let us calculate the position of Halley's comet at the beginning of the year 1986. In this case, the next (predicted) perihelion is given rather than the last (measured) perihelion so that we shall have to work backwards round the orbit from there.

Method	*Example*
The calculations for the comet	
1. Find the number of years since the epoch.	$Y = 1986.0–1986.112$
	$= -0.112$
2. Find $M_c = \dfrac{360\,Y}{T_p}$ degrees. Subtract multiples of 360 to bring the result back into the range 0–360, or if negative add 360.	$M_c = -0.530\,470$ degrees $+ 360$ $= 359.469\,530$ degrees
3. Convert M_c to radians by multiplying by $\dfrac{\pi}{180}$ ($\pi = 3.141\,592\,7$).	$M_c = 6.273\,927$ radians
4. Now we solve Kepler's equation by the routine R2 (§ 43). First guess (from Kepler's graphs):	$E_0 = M + (-0.2)$ $= 6.073\,927$ radians
5. Find a more accurate solution using R2.	$E = 6.056\,978$ radians
6. Calculate $\tan\dfrac{\nu}{2} = \left[\dfrac{1+e}{1-e}\right]^{\frac{1}{2}} \tan\dfrac{E}{2}$ (all angles in radians).	$\tan\dfrac{\nu}{2} = -0.881\,041$
7. Take inverse tan and double to get ν.	$\nu = -1.444\,483$ radians
8. Convert to degrees by multiplying by $\dfrac{180}{\pi}$ ($\pi = 3.141\,592\,7$).	$\nu = -82.762\,753$ degrees
9. Find $l = \nu + \varpi$.	$l = 87.248\,247$ degrees
10. Find $r = \dfrac{a(1-e^2)}{1+e\cos\nu}$.	$r = 1.028\,934$ AU
11. Calculate $\psi = \sin^{-1}\{\sin(l - \Omega)\sin i\}$.	$\psi = 8.530\,351$ degrees

Method (*continued*)	*Example*
12. Find $y = \sin(l - \Omega)\cos i$.	$y = -0.463\,070$
13. Find $x = \cos(l - \Omega)$.	$x = 0.873\,821$
14. Find $\tan^{-1}\left\{\frac{y}{x}\right\}$ and remove the ambiguity by comparing the signs of x and y with Figure 10. If not in the correct quadrant, add or subtract 180 or 360.	$\tan^{-1}\left\{\frac{y}{x}\right\} = -27.920\,776$ degrees
15. Add Ω to find l'.	$l' = 30.233\,224$ degrees
16. Find $r' = r\cos\psi$.	$r' = 1.017\,551$ AU
Now do the calculations for the Earth	
17. Find $N_E = \frac{360}{365.2422} \times \frac{D}{T_E}$, where D is the number of days since 1980.0 (see Table 2). Subtract multiples of 360 to bring the result into the range 0–360.	$D = 2192$ $N_E = 2160.452\,534$ $= 0.452\,534$ degrees
18. Calculate $M_E = N_E + \varepsilon - \varpi$ (see Table 7).	$M_E = -3.310\,329$ degrees
19. Find $L = N_E + \frac{360}{\pi} e \sin(M_E) + \varepsilon$ ($\pi = 3.141\,592\,7$). If the result is more than 360, subtract 360. If the result is negative, add 360.	$L = 99.175\,451$ degrees
20. Calculate $\nu_E = L - \varpi$.	$\nu_E = -3.420\,952$ degrees
21. Find $R = \frac{(1 - e^2)}{1 + e\cos\nu_E}$.	$R = 0.983\,311$ AU
22. If r' is less than R, calculate λ by using equation (a); if r' is more than R, use equation (b): (a) $\lambda = 180 + L + \tan^{-1}\left\{\frac{r'\sin(L - l')}{R - r'\cos(L - l')}\right\}$,	$r' > R$; use equation (b)
(b) $\lambda = \tan^{-1}\left\{\frac{R\sin(l' - L)}{r' - R\cos(l' - L)}\right\} + l'$. Add 360 to the result if negative.	$\lambda = 336.132\,193$ degrees
23. Find $\beta = \tan^{-1}\left\{\frac{r'\tan\psi\sin(\lambda - l')}{R\sin(l' - L)}\right\}$	$\beta = 7.673\,262$ degrees
24. Finally calculate the right ascension and declination using the method given in § 27.	$\alpha =$ **22h 20m** $\delta =$ **−2° 07′**

Thus we may expect to see the Halley's comet in the constellation of Aquarius at the beginning of 1986. How accurate our prediction is remains to be seen. We must not expect great precision, however, as the orbital elements change from return to return because of the perturbations to the comet's orbit by the gravitational fields of the planets. However, we can be confident that in this case the comet will be well within one degree of our estimated position.

58 Parabolic orbits

In preceding sections, we have calculated the orbits of members of the Solar System which are gravitationally bound to the Sun, like the planets and the periodic comets. These objects move in elliptical orbits with the Sun at a focus of the ellipse, and in the absence of perturbations from other members of the Solar System or from external influences, would continue to move indefinitely along the same elliptical paths. However, some comets do not seem to be bound to the Sun. If unperturbed they would appear once and shoot off into space never to return again. Their orbits are often defined in terms of *parabolic motion* and we have to use a slightly different procedure for calculating their positions, given the parabolic orbital elements:

t_0 = the epoch of the perihelion
q = perihelion distance in AU
i = inclination of the orbit
ω = argument of the perihelion ($\omega = \varpi - \Omega$)
Ω = longitude of the ascending node

The calculations proceed on much the same lines as for an elliptical orbit; once we have found the true anomaly, ν, and radius vector, r, we can use exactly the same method as in section 57 to calculate the position of the comet. However, the calculation of ν and r is slightly different.

First, we find the value of the quantity

$$W = \frac{0.036\,491\,162\,4}{q\sqrt{q}} \times d$$

where d is the number of days since the comet passed through perihelion. Next we have to solve equation

$$s^3 + 3s - W = 0.$$

This is best done by means of the iterative method shown in routine R3. Finally, calculate ν and r from

$$\nu = 2\tan^{-1}\{s\},$$
$$r = q(1 + s^2).$$

For example, the International Astronomical Union issued the following data on comet Kohler (1977m; IAUC 3137):

$t = 1977$ November 10.5659
$q = 0.990\,662$
$i = 48.7196$ degrees
$\omega = 163.4799$ degrees
$\Omega = 181.8175$ degrees

(hence $\varpi = \omega + \Omega = 345.2974$ degrees).

Routine R3. To solve the equation $s^3 + 3s - W = 0$.

1. First guess put $s = s_0 = \dfrac{W}{3}$.
2. Calculate $\delta = s^3 + 3s - W$.
3. If $|\delta| < \varepsilon$, go to step 6.
   Otherwise proceed with step 4.
   ε is the required accuracy (e.g. 10^{-6} degree).
4. Calculate $s_1 = \dfrac{2s^3 + W}{3(s^2 + 1)}$
   (note that s^3 can be calculated from $s \times s^2$).
5. Set $s = s_1$ and go to step 2.
6. The current value of s is within $\pm\varepsilon$ of the correct value.

The values of i, ω and Ω quoted here were referred to the standard equinox of 1950.0. Strictly, we should refer all our calculations to the same equinox, but we shall ignore the small error introduced by not doing so. Let us calculate the position of the comet on Christmas Day 1977, assuming no perturbations to its orbit.

Method	*Example*
1. Calculate the number of days since the epoch of perihelion. This can be done (for example) by subtracting the Julian date (§ 4) of the epoch from the Julian date of the day in question.	Nov. 10th 1977: $JD_1 = 2\,443\,457.5$ Dec. 25th 1977: $JD_2 = 2\,443\,502.5$ $JD_2 - JD_1 =$ 45 days Epoch = November 10.5659 $\therefore d =$ 44.4341 days
2. Find $W = \dfrac{0.036\,491\,162\,4}{q\sqrt{q}} \times d$.	$W =$ 1.644 432
3. Solve $s^3 + 3s - W = 0$ by using routine R3.	First guess $s = 0.548\,144$ $s =$ 0.505 171
4. Find $\nu = 2\tan^{-1}\{s\}$ and $r = q(1+s^2)$.	$\nu =$ 53.603 189 degrees $r =$ 1.243 477 AU
5. Carry on at instruction 9 (ignoring instruction 10) of § 57 to find α and δ.	$l =$ 398.900 589 degrees $\psi =$ −26.944 536 degrees $l' =$ 28.320 864 degrees $r' =$ 1.108 492 AU *For the Earth* $D =$ −736 days $N_E =$ −725.407 420 degrees = 354.592 580 degrees $L =$ 93.120 810 degrees $R =$ 0.983 503 AU $r' > R$; use equation (b) $\lambda =$ 336.099 109 degrees $\beta =$ −26.585 505 degrees
Here is the result:	$\alpha =$ **23h 17m** $\delta =$ **−33° 41′**

Comet Kohler was in the constellation of Sculptor on Christmas Day 1977.

59 Binary-star orbits

Quite often an astronomer sees a pair of stars very close together in his telescope. This apparent closeness may be just because two quite unrelated stars happen to lie near to the same line of sight. Sometimes, however, the stars are actually close to one another in space and they may then form a binary star in which each is bound to the other by mutual gravitational attraction. The stars describe elliptical orbits about one another, just as Jupiter describes an elliptical orbit about the Sun. The brighter of the two stars is generally called the *primary* and the fainter is called the *companion*; we shall consider that the companion orbits about the primary which is fixed in space, although really both stars orbit about their common centre of mass.

Figure 30 shows the appearance of a binary star. A is the primary, B the companion, and the line NAS is the observer's meridian through A; AN therefore defines the direction north. The line joining A to B is at position-angle θ (measured eastwards as shown) and of length ρ. Provided that we know

Figure 30. A binary star seen from the Earth.

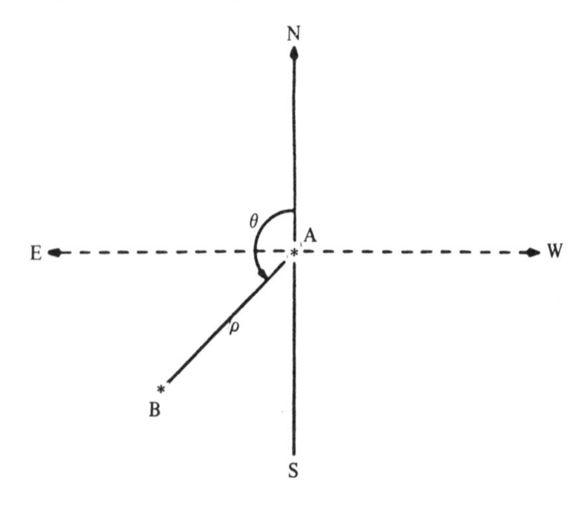

the elements of the binary orbit, we can calculate the values of θ and ρ and hence we can predict the appearance of the binary star at any time.

A binary-star orbit is drawn in Figure 31. The sphere is centred on the primary star, A, and its companion, B, describes an orbit about it shown by the small shaded ellipse in the centre. The great circle NL′DM′ shows where the plane through A perpendicular to the line of sight cuts the sphere. This plane is the plane of the sky as seen from the Earth. The line AN defines the direction north as in Figure 30. The great circle L′P′B′M′ shows where the plane of the true binary orbit cuts the sphere. The point L′ is the projection of the ascending node, L, onto the sphere, M′ the projection of the descending

Figure 31. A binary-star orbit.

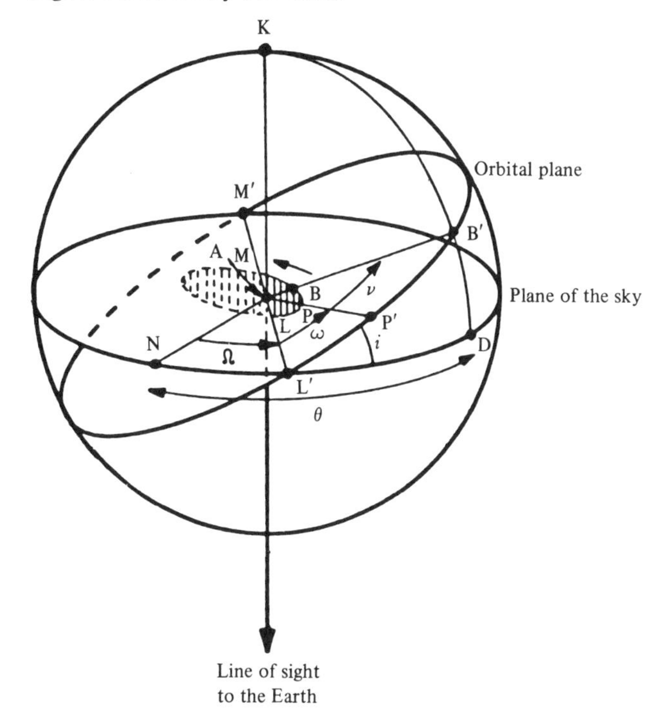

node, M, and P′ the projection of the point of closest approach, P, the *periastron*. The companion star is at B. Longitudes are reckoned from the ascending node, L, and the true anomaly, ν, is the angle between B and the periastron. We need the following elements to calculate the orbit:

T = the period of revolution
t = the epoch of periastron
e = the eccentricity of the orbit
a = the semi-major axis of the orbit
i = the inclination of the orbit to the plane of the sky
Ω = the position-angle of the ascending node
ω = the longitude of the periastron.

All angles are measured in the direction of motion. The elements for some binary stars are listed in Table 9.

The calculation of a binary-star orbit proceeds in much the same way as that of a planetary orbit. We first find the mean anomaly, M, from

$$M = \frac{360\,Y}{T},$$

where Y is the number of years since the epoch of periastron. Next we must solve Kepler's equation

$$E - e \sin E = M,$$

using the method given in section 57. The true anomaly, ν, and radius vector, r, can then be found from

$$\nu = 2 \tan^{-1}\left\{\left[\frac{1+e}{1-e}\right]^{\frac{1}{2}} \tan \frac{E}{2}\right\}$$

and

$$r = a(1 - e \cos E)$$

(remembering that E has been found in radians). Finally, θ is given by

$$\theta = \tan^{-1}\left\{\frac{\sin(\nu+\omega)\cos i}{\cos(\nu+\omega)}\right\} + \Omega \text{ degrees},$$

Table 9. The orbital elements of some binary stars

Name	Period, T (mean solar years)	Epoch of periastron, t	Longitude of periastron, ω (degrees)	Eccentricity, e	Semi-major axis of orbit, a (arcsec)	Inclination of orbit, i (degrees)	Position-angle of ascending node, Ω (degrees)
η Coronae Borealis	41.623	1934.008	219.907	0.2763	0.907	59.025	23.717
γ Virginis	171.37	1836.433	252.88	0.8808	3.746	146.05	31.78
η Cassiopeiae	480	1889.6	268.59	0.497	11.9939	34.76	278.42
ζ Orionis	1508.6	2070.6	47.3	0.07	2.728	72.0	155.5
α Canis Majoris (Sirius)	50.09	1894.13	147.27	0.5923	7.500	136.53	44.57
δ Geminorum	1200	1437	57.19	0.1100	6.9753	63.28	18.38
α Geminorum (Castor)	420.07	1965.3	261.43	0.33	6.295	115.94	40.47
α Canis Minoris (Procyon)	40.65	1927.6	269.8	0.40	4.548	35.7	284.3
α Centauri	79.920	1955.56	231.560	0.516	17.583	79.240	204.868
α Scorpionis (Antares)	900	1889.0	0.0	0.0	3.21	86.3	273.0

and ρ from

$$\rho = \frac{r \cos(\nu + \omega)}{\cos(\theta - \Omega)} \text{ degrees.}$$

For example, let us calculate the visual aspect of the binary system η Coronae Borealis at the beginning of 1980.

Method	*Example*
1. Find the number of years since the epoch.	$Y = 1980.0 - 1934.008$ $= 45.992$ years
2. Find $M = \frac{360Y}{T}$. Subtract multiples of 360 to bring the result into the range 0–360.	$M = 397.787\,762$ degrees $= 37.787\,762$ degrees
3. Convert to radians by multiplying by $\frac{\pi}{180}$ ($\pi = 3.141\,592\,7$).	$M = 0.659\,521$ radians
4. Solve Kepler's equation $E - e \sin E = M$ by the method outlined in § 57.	First guess $E_0 = 0.86$ radians Solution is $E = 0.870\,858$ radians
5. Find $\nu = 2 \tan^{-1}\left\{\left[\frac{1+e}{1-e}\right]^{\frac{1}{2}} \tan\frac{E}{2}\right\}$, all angles in radians.	$\nu = 1.106\,803$ radians
6. Multiply by $\frac{180}{\pi}$ to convert to degrees ($\pi = 3.141\,592\,7$).	$\nu = 63.415\,137$ degrees
7. Find $r = a(1 - e \cos E)$, remembering that E is expressed in radians.	$r = 0.745\,568$ arcsec
8. Calculate $y = \sin(\nu + \omega) \cos i$.	$y = -0.500\,814$
9. Calculate $x = \cos(\nu + \omega)$.	$x = 0.230\,426$
10. Find $\tan^{-1}\left\{\frac{y}{x}\right\}$ and remove the ambiguity by reference to Figure 10. Add or subtract 180 or 360 to bring the result into the correct quadrant, unless it is already in the correct quadrant.	$\tan^{-1}\left\{\frac{y}{x}\right\} = -65.292\,744$ $+ 360$ $= 294.707\,256$ degrees
11. Add Ω to find θ. Subtract 360 if more than 360. Add 360 if negative.	$\theta =$ **318.424 degrees**
12. Find $\rho = \frac{r \cos(\nu + \omega)}{\cos(\theta - \Omega)}$.	$\rho =$ **0.411 arcsec**

The Moon and eclipses

Of all the heavenly bodies visible in the night from the Earth, the Moon is the most spectacular. It far outshines even the most brilliant planet, moves so quickly that you can see its motion against the stars, and provides a wealth of detail in the shadowy features of its disc. Yet its motion is the most difficult to predict and it is for that reason we have left it until last. It is of course in orbit about the Earth but the Sun and other members of the Solar System perturb that orbit to such an extent that many corrections are needed to calculate the Moon's position accurately.

In the next few sections we use a simple method to find the position of the Moon. The method takes account only of the principle perturbations to the orbit yet gives results which are accurate enough for most purposes. We also calculate the times of moonrise and moonset, the phases of the Moon, and the circumstances of both solar and lunar eclipses. The calculations are lengthy but the satisfaction you feel when you accurately predict, for example, the occurrence of a lunar eclipse, cannot be denied.

60 The Moon's orbit

To an Earth-bound observer, the Moon appears to be in orbit about the Earth, making one complete revolution with respect to the background of stars in 27.3217 days. This period is called the *sidereal month*. During this time the Earth moves on along its own orbit so that the Sun's position changes with respect to the stars. Hence the Moon has some extra distance to make up to regain its position relative to the Sun. The interval defined by the time taken for the Moon to return to the same position relative to the Sun is called the *synodic month* and is equal to 29.5306 days. The direction of motion of the Moon in its orbit about the Earth is *prograde*; that is, it is in the same sense as that of all the planets about the Sun.

A celestial observer viewing the Solar System from a great distance would not, however, see the Moon making loops in space about the Earth. Rather, he would describe the situation by saying that the Moon is in orbit around the Sun, as is the Earth, and that the effect of the Earth's influence is to make the Moon's orbit wiggle a little as the relative positions of Earth and Moon change (Figure 32). This is because the Sun's gravitational force on the Moon is much greater than that of

Figure 32. The Moon's orbit (not to scale).

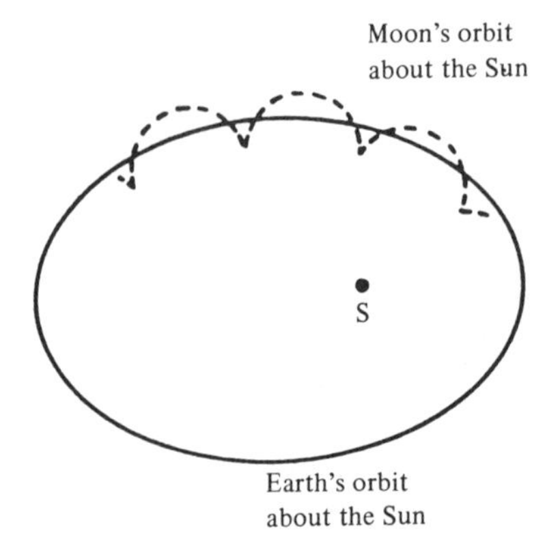

the Earth, even though the latter is nearer. It is hardly surprising that the orbit of the Moon is so difficult to calculate since it is regulated by two bodies, not one, and the two bodies are themselves tied in orbit about each other.

For the purposes of our calculations, we are going to imagine that both the Sun and the Moon are in orbit about the Earth. We have already calculated the position of the Sun by this means in section 42. We will need those calculations in the next few sections to find the magnitude of some of the corrections to the Moon's orbit.

There are three main effects of the perturbations caused by the Sun on the Moon's apparent orbit round the Earth. The first of these is called *evection* in which the apparent value of the eccentricity of the Moon's orbit varies slightly. The second is due to the variation of the Earth–Sun distance as the Earth travels in its own ellipse about the Sun. This correction is called the *annual equation*. The third inequality takes account of the motion of the Moon in the Sun's gravitational field. When the Moon is on one side of the Earth it is nearer the Sun so that the Sun's gravitational attraction is slightly more than when the Moon is on the other side of the Earth. This correction is called the *variation*.

These corrections alone, together with the usual correction called the equation of the centre, can make up to 9° of difference in the Moon's mean anomaly so it is important that they be taken into account. We shall make six corrections in all to find the position of the Moon to within one fifth of a degree.

The apparent motions of the Moon and the Sun about the Earth are drawn in Figure 33. This diagram is similar to that of Figure 21 except that here the Earth is at the centre and both the Sun and the Moon describe ellipses about the Earth. Once again you are to imagine that you are looking at the Solar System from a great distance and, further, that you are moving in such a manner that the Earth appears to be stationary in your view. The large sphere is centred on the Earth, E, and the planes of the orbits of Sun and Moon are projected to cut the sphere along the circles ♈$N'_1S'N'_2$ and $N'_1P'm'N'_2$ respectively. S′ is the projection of the Sun onto this sphere and its longitude, measured from the first point of Aries, ♈, is denoted by $\lambda_\odot$. The Moon's orbit is inclined to the ecliptic at an angle i; N'_1 and N'_2 are the projections of the ascending and descending nodes, P′ is the projection of the Moon's perigee, and m′ is the projection of the present position of the Moon. The longitude of the ascending node is ☊, the longitude of the perigee is ☊ + ω and the Moon's true anomaly is ν.

There are two principal effects of the perturbations mentioned above. The first is that the perigee of the Moon's orbit, unlike the (nearly) stationary perihelia of the planet's orbits, advances (prograde) at such a rate that it makes one complete revolution in 8.85 years. The second is that the line joining the nodes, $N'_1N'_2$ moves backwards (retrograde) around the ecliptic so that it makes one complete revolution in 18.61 years. Yet another month can be defined by the time it takes the Moon to return to its ascending node. This is the *draconic* or *nodal month* and it is equal to 27.2122 days.

Figure 33. Defining the Moon's orbit.

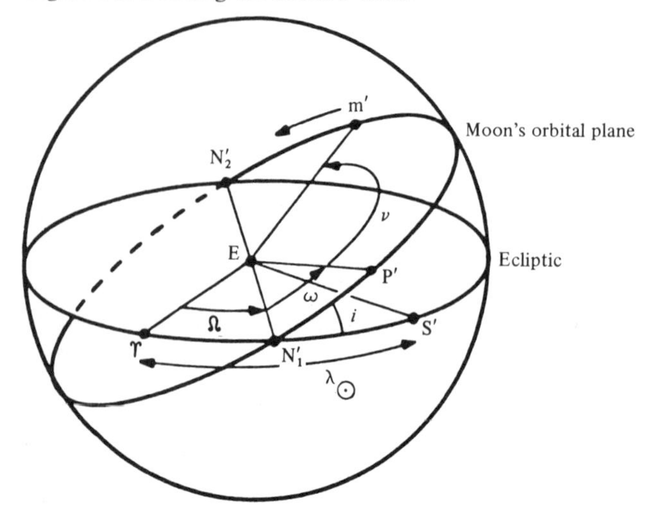

61 Calculating the Moon's position

The steps involved in the process of finding the position of the Moon are much the same as those involved in calculating the position of a planet, except that (i) correction terms have to be applied at every step and (ii) the longitudes of the ascending node and perigee cannot be regarded as constant. We first determine the Moon's mean anomaly, M_m, which refers to the position of a fictitious Moon in uniform circular motion about the Earth. Then we find the longitude, and, by referring it to the plane of the ecliptic, the geocentric ecliptic coordinates λ_m and β_m. Finally, we convert to right ascension and declination using the method given in section 27.

Once again we choose the epoch 1980 January 0.0 as starting point. We calculate the number of days, D, since the epoch to the required date and time, counting the time of day as a fraction of a day. For slightly better accuracy we should use the ephemeris time, ET, rather than the universal time, UT (see section 16). Then we find:

(*a*) the Sun's ecliptic longitude, $\lambda_\odot$, and mean anomaly, $M_\odot$, by the method given in section 42;

(*b*) the Moon's mean longitude, l, given by

$$l = 13.176\,396\,6D + l_0;$$

(*c*) the Moon's mean anomaly, M_m, given by

$$M_m = l - 0.111\,404\,1D - P_0;$$

(*d*) the ascending node's mean longitude, N, given by

$$N = N_0 - 0.052\,953\,9D.$$

l_0, P_0 and N_0 are the mean longitudes at the epoch.

Next we calculate the corrections for evection, E_v, the annual equation, A_e, and a third correction, A_3:

$$E_v = 1.2739 \sin[2C - M_m],$$

$$A_e = 0.1858 \sin(M_\odot),$$

$$A_3 = 0.37 \sin(M_\odot),$$

where $C = l - \lambda_\odot$. With these corrections we can find the Moon's corrected anomaly, M'_m:

$$M'_m = M_m + E_v - A_e - A_3.$$

We can now find the correction for the equation of the centre:

$$E_c = 6.2886 \sin(M'_m).$$

Yet another correction term must be calculated:

$$A_4 = 0.214 \sin(2M'_m).$$

Now we can find the value of the Moon's corrected longitude, l', from

$$l' = l + E_v + E_c - A_e + A_4.$$

The final correction to apply to the Moon's longitude is the variation, V, given by

$$V = 0.6583 \sin 2(l' - \lambda_\odot).$$

Table 10. Elements of the Moon's orbit, epoch 1980.0

Moon's mean longitude at the epoch:	$l_0 =$	64.975 464 degrees
Mean longitude of the perigee at the epoch:	$P_0 =$	349.383 063 degrees
Mean longitude of the node at the epoch:	$N_0 =$	151.950 429 degrees
Inclination of the Moon's orbit:	$i =$	5.145 396 degrees
Eccentricity of Moon's orbit:	$e =$	0.054 900
Moon's angular size at distance a from the Earth:	$\theta_0 =$	0.5181 degrees
Semi-major axis of Moon's orbit	$a =$	384 401 km
Parallax at distance a from the earth	$\pi_0 =$	0.950 7 degrees

Then the Moon's true orbital longitude, l'', is just

$$l'' = l' + V.$$

Referring the longitude to the ecliptic allows us to calculate the ecliptic latitude, β_m, and longitude, λ_m. Thus

$$\lambda_m = \tan^{-1}\left\{\frac{\sin(l''-N')\cos i}{\cos(l''-N')}\right\} + N',$$

and

$$\beta_m = \sin^{-1}\{\sin(l''-N')\sin i\},$$

where N' is the corrected longitude of the node, and it is given by

$$N' = N - 0.16\sin(M_\odot).$$

This is a lengthy calculation! Let us illustrate it with an example: what was the position of the Moon on February 26th 1979 at 16h 00m UT? The values of l_0, P_0, N_0 and other parameters of the Moon's orbit are listed in Table 10.

Method	*Example*
1. Find the number of days, d, since January 0.0 (§ 3). Remember to count the hours, minutes and seconds as a fraction of a day.	16h 00m UT = 16h 00m 50s ET (§ 16) = 16.013 889 hours ÷ 24 = 0.667 245 days + 57 d = 57.667 245 days
2. Add 365 days for each year since 1980 plus one for each leap year (see Table 2). The total is D.	− 365.0 D = −307.332 755 days
3. Find $\lambda_\odot$ and $M_\odot$ using the method of § 42.	$\lambda_\odot$ = 337.448 134 degrees $M_\odot$ = 53.315 427 degrees
4. Find $l = 13.176\,396\,6D + l_0$. Adjust to the range 0–360 by adding or subtracting multiples of 360.	l = 335.437 196 degrees
5. Find $M_m = l - 0.111\,404\,1D - P_0$. Adjust to the range 0–360.	M_m = 20.292 231 degrees
6. Find $N = N_0 - 0.052\,953\,9D$. Adjust to the range 0–360.	N = 168.224 897 degrees

Method (continued)	*Example*
7. Calculate $E_v = 1.2739 \sin[2C - M_m]$ where $C = l - \lambda_\odot$.	$E_v = -0.524\ 514$ degrees
8. Find $A_e = 0.1858 \sin(M_\odot)$ and $A_3 = 0.37 \sin(M_\odot)$.	$A_e = 0.149\ 000$ degrees $A_3 = 0.296\ 717$ degrees
9. Find the corrected anomaly: $M'_m = M_m + E_v - A_e - A_3$.	$M'_m = 19.322\ 001$ degrees
10. Calculate $E_c = 6.2886 \sin(M'_m)$.	$E_c = 2.080\ 752$ degrees
11. Calculate $A_4 = 0.214 \sin(2M'_m)$.	$A_4 = 0.133\ 639$ degrees
12. Find $l' = l + E_v + E_c - A_e + A_4$.	$l' = 336.978\ 073$ degrees
13. Find $V = 0.6583 \sin[2(l' - \lambda_\odot)]$.	$V = -0.010\ 801$ degrees
14. Hence find the true longitude $l'' = l' + V$.	$l'' = 336.967\ 272$ degrees
15. Find $N' = N - 0.16 \sin(M_\odot)$.	$N' = 168.096\ 587$ degrees
16. Find $y = \sin(l'' - N') \cos i$.	$y = 0.192\ 246$
17. Find $x = \cos(l'' - N')$.	$x = -0.981\ 194$
18. Calculate $\tan^{-1}\left\{\frac{y}{x}\right\}$. Remove the ambiguity by reference to Figure 10, adding or subtracting 180 or 360 to bring the result into the correct quadrant unless it is already there.	$\tan^{-1}\left\{\frac{y}{x}\right\} = -11.085\ 570$ $+180$ $= 168.914\ 430$
19. Add N' to find λ_m.	$\lambda_m = 337.011\ 006$ degrees
20. Find $\beta_m = \sin^{-1}\{\sin(l'' - N') \sin i\}$.	$\beta_m = 0.991\ 900$ degrees
21. Convert to right ascension and declination by the method of § 27.	$\alpha_m =$ **22h 33m 27s** $\delta_m =$ **−8° 01′ 01″**

The *Astronomical Ephemeris* gives the apparent coordinates of the Moon at 16h 00m ET as $\alpha = 22\text{h}\ 33\text{m}\ 29\text{s}$ and $\delta = -8°\ 02'\ 42''$. We may expect an error of up to one fifth of a degree in ecliptic coordinates, as illustrated in Figure 34 where the error, Δ, between λ_m calculated by this method and that quoted in the *Astronomical Ephemeris* is drawn as a function of the date for early 1979.

62 The Moon's hourly motions

The calculation which we have to do to find the position of the Moon is a lengthy affair and needs great care in its execution to avoid making mistakes. It may be that you require the position at several different times during one day and, rather than repeat the calculation several times, it is sufficient to find the position once and then extrapolate to the other times using the values for the hourly motions of the Moon in ecliptic latitude and longitude. These motions are given by the formulae

$$\Delta\beta = 0.05 \cos(l'' - N') \text{ degrees/hour},$$

$$\Delta\lambda = 0.55 + 0.06 \cos(M'_m) \text{ degrees/hour},$$

where $\Delta\beta$ is the motion in latitude and $\Delta\lambda$ is the motion in longitude. Given a position λ_0, β_0 at time t_0, the position after t hours is simply

$$\beta = \beta_0 + \Delta\beta t,$$

$$\lambda = \lambda_0 + \Delta\lambda t.$$

Figure 34. The error, $\Delta = \lambda_m(\text{DS}) - \lambda_m(\text{AE})$, between the ecliptic coordinates of the Moon as calculated by the method given here (λ_m(DS)) and those quoted in the *Astronomical Ephemeris* (λ_m(AE)), for early 1979.

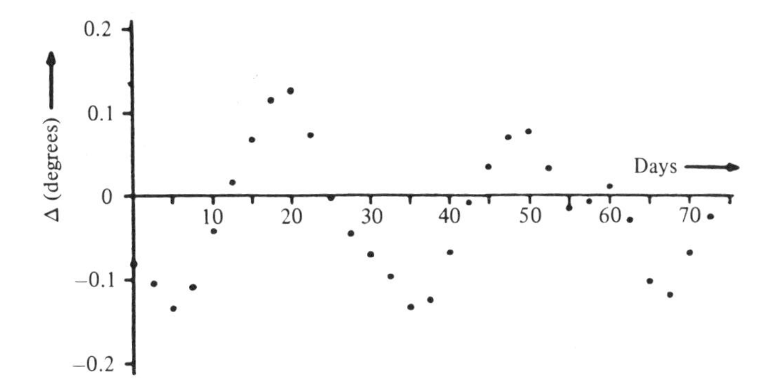

Continuing the previous example, what were the Moon's ecliptic coordinates at 17h 00m UT on February 26th 1979?

Method	*Example*
1. Write down λ_0, l''_0, β_0, N', M'_m, t_0 (§ 61).	$\lambda_0 =$ 337.011 006 degrees $l''_0 =$ 336.967 272 degrees $\beta_0 =$ 0.991 990 degrees $N' =$ 168.096 587 degrees $M'_m =$ 19.322 001 degrees $t_0 =$ 16.00 hours
2. Calculate $\Delta\beta = 0.05 \cos(l''_0 - N')$ and $\Delta\lambda = 0.55 + 0.06 \cos(M'_m)$.	$\Delta\beta =$ −0.049 060 degrees/hour $\Delta\lambda =$ 0.606 620 degrees/hour
3. Find t in hours: $t =$ new time $- t_0$. Don't forget to express both times in decimal hours (§ 7).	$t = 17 - 16$ $= 1$ hour
4. Find the new coordinates: $\beta = \beta_0 + \Delta\beta t$, $\lambda = \lambda_0 + \Delta\lambda t$.	 $\beta =$ **0.943 degrees** $\lambda =$ **337.618 degrees**

63 The phases of the Moon

The relative positions of the Sun and the Moon as viewed from the Earth change during the course of one month. It is always the hemisphere of the Moon facing towards the Sun which is brightly illuminated but we on the Earth see only that half which faces us. Unless the Moon is in opposition to the Sun, the time of full Moon, our half is not uniformly illuminated but overlaps both the bright and dark sides; hence we see only a segment of the disc. The area of the segment expressed as a fraction of the whole disc is called the phase.

The variation of phase with the Moon's position is illustrated in Figure 35. Here a plan view of the Moon's orbit about the Earth, E, is shown. The Moon is shown in five positions marked 1 to 5. At 1, the whole of the dark side is turned towards us so that unless the Moon is illuminated by sufficient *earthshine* it is invisible. This is the new Moon. One

Figure 35. The phases of the Moon.

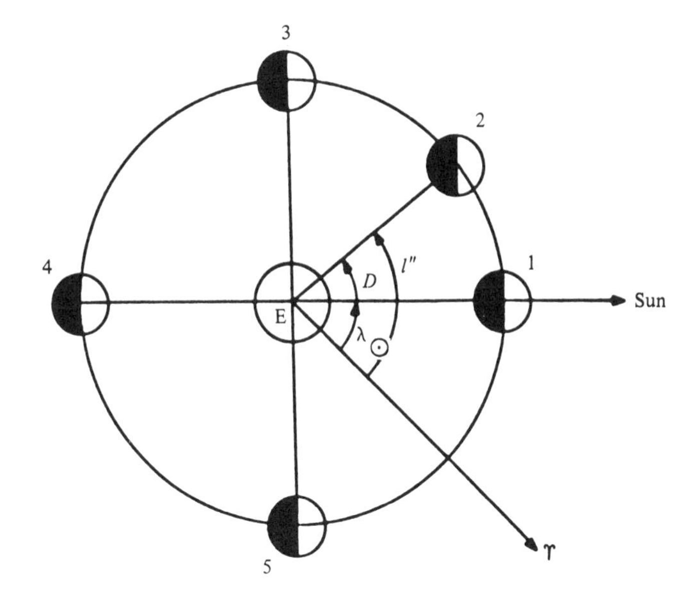

week later the Moon has reached position 3 and is said to be *in quadrature*. This is the first quarter. Position 4 is the full Moon, the point of opposition with the Sun. At position 5, the Moon is again in quadrature; this is the third quarter. Between positions 3 and 5 more than half of the Moon's face is illuminated and the Moon is said to be *gibbous*.

In Figure 35, the angle D is called the *age* of the Moon, varying from 0° to 360° as the Moon completes one cycle of its orbit. Sometimes this angle is expressed in days, one day being equivalent to about 13°. The phase, F, is given by

$$F = \tfrac{1}{2}(1 - \cos D).$$

We have already made most of the calculations to find D in section 61. Referring again to Figure 35, we find

$$D = l'' - \lambda_{\odot} \text{ degrees.}$$

Continuing the example of section 61, we will find the phase of the Moon at 16h 00m UT on February 26th 1979.

Method	*Example*
1. Find the values of $\lambda_{\odot}$ and l'', using the method of § 61.	$\lambda_{\odot} = 337.448\,134$ degrees $l'' = 336.967\,272$ degrees
2. Find $D = l'' - \lambda_{\odot}$.	$D = -0.480\,862$ degrees
3. Calculate $F = \tfrac{1}{2}(1 - \cos D)$.	$F =$ **0.0**

We see that this is near the moment of new Moon; it is also near the moment of a total solar eclipse which we shall calculate in section 70.

64 The position-angle of the Moon's bright limb

In section 55 we saw how to calculate the position-angle, χ, of the bright limb of a planet. χ is defined to be the angle of the midpoint of the illuminated limb measured eastwards from the north point of the disc (see Figure 27). We can do the same for the Moon. χ is given by

$$\chi = \tan^{-1}\left\{\frac{\cos\delta_\odot \sin(\alpha_\odot - \alpha_m)}{\cos\delta_m \sin\delta_\odot - \sin\delta_m \cos\delta_\odot \cos(\alpha_\odot - \alpha_m)}\right\},$$

where $\alpha_\odot$, $\delta_\odot$ and α_m, δ_m are the equatorial coordinates of the Sun and Moon respectively.

For example, what was the position-angle of the Moon's bright limb on May 19th 1979? The coordinates of Sun and Moon that day were $\alpha_\odot = 03\text{h}\,40\text{m}\,38\text{s}$, $\delta_\odot = 19°\,35'\,16''$, $\alpha_m = 21\text{h}\,56\text{m}\,32\text{s}$, and $\delta_m = -10°\,57'\,08''$. (These can be calculated by the methods given in sections 42 and 61.)

Method	*Example*
1. Convert $\alpha_\odot$ and α_m first to decimal hours (§ 7) and then to degrees (§ 22).	$\alpha_\odot =$ 3.677 222 hours
	$=$ 55.158 333 degrees
	$\alpha_m =$ 21.942 222 hours
	$=$ 329.133 333 degrees
2. Convert $\delta_\odot$ and δ_m to decimal degrees (§ 21).	$\delta_\odot =$ 19.587 778 degrees
	$\delta_m =$ −10.952 222 degrees
3. Find $y = \cos\delta_\odot \sin(\alpha_\odot - \alpha_m)$.	$y =$ 0.939 863
4. Find $x = \cos\delta_m \sin\delta_\odot - \sin\delta_m \cos\delta_\odot \cos(\alpha_\odot - \alpha_m)$.	$x =$ 0.341 553
5. Find $\chi' = \tan^{-1}\left\{\frac{y}{x}\right\}$. Remove the ambiguity by referring to Figure 10. If χ' is not in the correct quadrant, add or subtract 180 or 360 to make it so. Otherwise $\chi = \chi'$.	$\chi' =$ 70.028 516 degrees
	$\chi =$ **70.03 degrees**

65 The Moon's distance, angular size and horizontal parallax

During the course of one complete circuit of its orbit, the Moon's distance, ρ, from the Earth varies quite considerably. Its point of closest approach, the perigee, is about 356 000 km from the Earth while the furthest point, the apogee, is at a distance of 407 000 km. We can calculate its distance at any other point quite easily, as it is given by the formula

$$\rho = \frac{a(1-e^2)}{1+e\cos(M'_{\mathrm{m}}+E_{\mathrm{c}})},$$

where M'_{m} is the corrected anomaly, E_{c} is the correction for the equation of the centre (defined in section 61), e is the eccentricity and a the semi-major axis of the Moon's orbit. We usually wish to express the distance as a fraction of a so that we write

$$\rho' = \frac{\rho}{a} = \frac{(1-e^2)}{1+e\cos(M'_{\mathrm{m}}+E_{\mathrm{c}})}.$$

The units of ρ are the same as those of a; if a is expressed in kilometres, so is ρ.

The Moon's apparent angular diameter, θ, follows directly from the value of ρ'. It is given by

$$\theta = \frac{\theta_0}{\rho'},$$

where θ_0 is the Moon's apparent angular diameter when it is at a distance a from the Earth. The value of θ_0 is given in Table 10.

The Moon's horizontal parallax is defined to be the angle subtended at the Moon by the Earth's radius. In Figure 36 it is given by the symbol π (not to be confused with the constant 3.141 592 7). The formula is

$$\pi = \frac{\pi_0}{\rho'},$$

where π_0 is the horizontal parallax at distance a from the Earth (Table 10).

For example, what were the values of ρ', θ and π on September 6th 1979 at 0h UT?

Method	*Example*
1. Find M'_m and E_c by the method of § 61.	$M'_m = -359.735\,278$ degrees $E_c = 0.029\,055$ degrees
2. Calculate $\rho' = \dfrac{1-e^2}{1+e\cos(M'_m+E_c)}$ (Table 10 for e).	$\rho' =$ **0.945 101**
3. Find $\theta = \dfrac{\theta_0}{\rho'}$.	$\theta = 0.548$ degrees = **0° 32′ 54″**
4. Find $\pi = \dfrac{\pi_0}{\rho'}$.	$\pi = 1.005\,925$ = **1° 00′ 21″**

Figure 36. Lunar parallax.

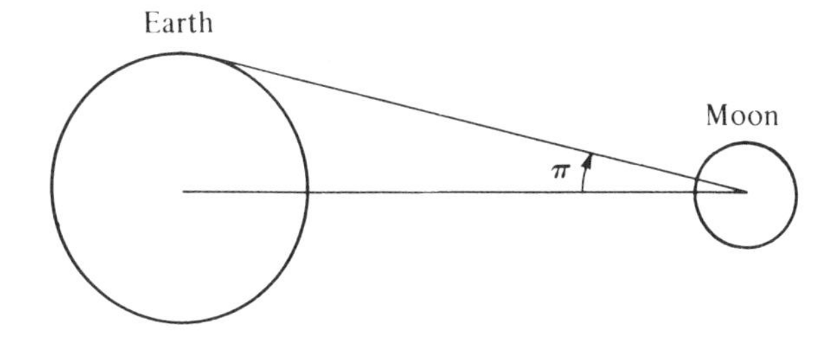

66 Moonrise and moonset

In section 32 we found how to calculate the times of rising or setting of a star given its equatorial coordinates. We can apply the same method to the Moon to find the times of moonrise and moonset but the problem is complicated by two factors. One is that the Moon is in rapid motion so that its right ascension and declination are continually changing. To find the time of moonset, for instance, we require the coordinates of the Moon at that time; but to find these coordinates we need the time of moonset, and so we go round in a circle. One way of overcoming the problem is to find the Moon's coordinates at two different times on the same day and interpolate between them to determine the required time; we adopt this method here.

The other major complication with the Moon is that it is, astronomically speaking, very close to the Earth. The coordinates which we work out are correct for the centre of the Earth, but when we observe from the Earth's surface the apparent coordinates change slightly; this effect is called *parallax* (see section 35). In the case of the Moon the parallax can be as much as a whole degree. We can calculate the effect of parallax by the method given in section 36. Taking this into account, together with the corrections for atmospheric refraction and the finite size of the Moon's disc (times quoted are for the upper limb), we can find the times of moonrise and moonset correct to within a few minutes of time.

We first calculate the Moon's right ascension and declination, α_1 and δ_1, at the midnight before the day in question using the method of section 61. Then we use the Moon's hourly motions in latitude and longitude (section 62) to find the coordinates, α_2 and δ_2, 12 hours later (midday on that day). Next we must apply the corrections for parallax to both sets of coordinates (section 36). The value of r can be found via the value of ρ', calculated in section 65, as follows:

$$r = 60.268\,322\rho'.$$

The corrected sets of coordinates, α'_1, δ'_1 and α'_2, δ'_2, can then be used to find the local sidereal times of rising and setting correct for the Moon's positions at midnight and at midday; call the time (rising or setting) for midnight ST1 and for midday ST2. Then we substitute into the interpolation formula below to find better approximations to the sidereal times, T, of moonrise and moonset:

$$T = \frac{12.03 \times \text{ST1}}{12.03 + \text{ST1} - \text{ST2}} \text{ hours.}$$

Finally, we apply the corrections for refraction and for the finite size of the Moon's disc just as we did for the Sun in section 45.

Let us clarify all this with an example: what were the times of moonrise and moonset on September 6th 1979 as observed from a place at sea-level on longitude 0° and at latitude 52° N?

Method	*Example*
1. Calculate the right ascension and declination of the Moon at midnight (§ 61).	$\lambda_1 = 336.292\,039$ degrees $\beta_1 = 0.174\,853$ degrees $\alpha_1 = 22.532\,709$ hours $\delta_1 = -9.041\,689$ degrees
2. Find the coordinates 12 hours later (§ 62).	$\lambda_2 = 343.612\,031$ degrees $\beta_2 = -0.424\,799$ degrees $\alpha_2 = 23.004\,309$ hours $\delta_2 = -6.836\,710$ degrees
3. Correct for parallax. Use § 35 to find $\rho \cos \phi'$ and $\rho \sin \phi'$. Use § 65 to find ρ' and hence $r = 60.268\,322\rho'$. Find the approximate hour-angle, H, at rising or setting from $H = \cos^{-1}\{-\tan \phi \tan \delta'\}$ (where δ' is the average value of δ_1 and δ_2) and correct for parallax by the method of § 36.	$\rho \cos \phi' = 0.616\,945$ $\rho \sin \phi' = 0.784\,367$ $r = 56.959\,651$ Earth-radii $\delta' = \frac{\delta_1 + \delta_2}{2} = -7.939$ degrees $H = 79.718$ degrees $\Delta = 0.617\,711$ degrees $\Delta = 0.041\,181$ hours $\alpha'_1 = 22.491\,528$ hours $\alpha'_2 = 22.963\,128$ hours $\delta'_1 = -9.839\,231$ degrees $\delta'_2 = -7.631\,825$ degrees

Method (continued)	*Example*
4. Use the method given in § 32 to find the local sidereal times of rising and setting corresponding to the two sets of corrected coordinates.	$ST1_r =$ 17.346 579 hours $ST2_r =$ 17.621 481 hours $ST1_s =$ 3.636 477 hours $ST2_s =$ 4.304 775 hours
5. Use the following interpolation formula to find better approximations to ST_r and ST_s: $$T = \frac{12.03 \times ST1}{12.03 + ST1 - ST2}.$$	$T_r =$ 17.752 242 hours $T_s =$ 3.850 375 hours
6. Calculate the correction, Δt, due to refraction and the finite size of the Moon's disc (§ 32). $R = 34'$ at the horizon (§ 34) and θ can be found from § 65. Use the average value of δ_1' and δ_2' $\left(\delta'' = \frac{\delta_1' + \delta_2'}{2}\right)$.	$\delta'' =$ −8.735 528 degrees $\psi =$ 37.130 849 degrees $\theta =$ 0.548 degrees $R = 34' = 0.567$ degrees $x = R + \frac{\theta}{2} = 0.841$ degrees $y =$ 1.393 308 degrees $\Delta t = 338.3$ seconds
7. Divide Δt by 3600 to convert to hours; add to T_s and subtract from T_r.	$\Delta t =$ 0.094 hours $T_r' =$ 17.658 265 hours $T_s' =$ 3.944 352 hours
8. Finally, convert to GMT (§ 13). Note that in this case the longitude is 0° so that no conversion from LST to GST is necessary (§ 15).	$GMT_r =$ **18h 38m** $GMT_s =$ **04h 58m**

The *Astronomical Ephemeris* gives the time of moonrise as 18h 46m and of moonset as 05h 02m. We could improve on our result by taking more care with the corrections for parallax, using the correct values of H and δ rather than the approximations assumed in the example, and by making one iteration to find more accurate values for ST_r and ST_s using the new coordinates, not forgetting to apply accurate corrections for parallax, refraction and the Moon's finite angular size each time.

67 Eclipses

Both the Earth and the Moon cast long shadows into space. The Earth's shadow lies exactly in the plane of the ecliptic opposite the Sun whereas that of the Moon may be above or below the ecliptic depending on the position of the Moon (Figure 37). The shadows are always present but we are usually unaware of them since we cannot see them from the Earth. Occasionally however one of the bodies passes through the shadow of the other and then we observe an eclipse: if the Moon passes through the Earth's shadow it is an eclipse of the Moon, or a *lunar eclipse*; when the Moon casts its shadow upon the Earth, we see the Sun partially or totally obscured and it is then a *solar eclipse*.

An eclipse of the Moon can only happen at full Moon and an eclipse of the Sun at new Moon. We do not see an eclipse on every such occasion, however, since the Moon's orbit does not lie in the plane of the ecliptic. Only when the Moon is near one of its nodes can an eclipse occur.

A lunar eclipse begins with the *penumbral phase* when the Moon enters the penumbra of the Earth's shadow, and the Moon's disc becomes a little fainter. You probably wouldn't notice this unless you were looking for it. As the Moon enters the umbra the *partial phase* begins; when it has all moved inside the umbra the Sun's light is entirely cut off and the *total phase* begins. The only light reaching the Moon is then that refracted round the edges of the Earth, giving the Moon a

Figure 37. Shadows cast by the Moon and Earth.

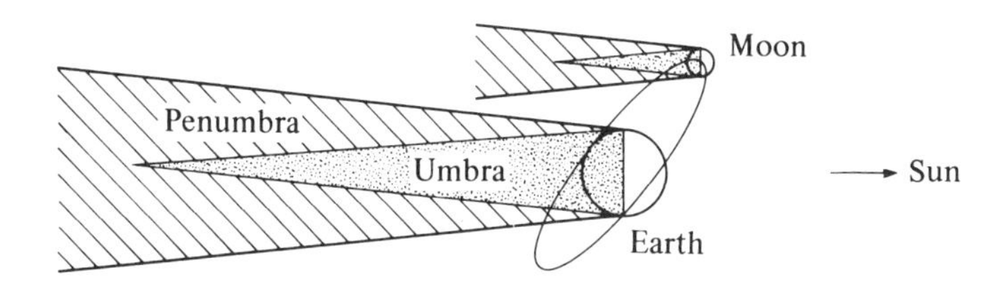

coppery hue. The Earth's shadow extends well beyond the Moon's orbit so that it is always possible for a total lunar eclipse to occur, other circumstances being favourable (Figure 38*a*).

A solar eclipse begins with the partial phase when the Earth enters the penumbra of the Moon's shadow. We see a 'bite' missing out of the Sun's disc and as the eclipse progresses the size of the bite increases. If you are favourably situated, you will see the Moon eventually cover the whole Sun. The eclipse is then total. Since the Moon is so much smaller than the Earth, its umbra extends a much shorter distance into space, in fact only just far enough to reach the Earth when the conditions are right (Figure 38*b*). The tip of the umbra casts a small shadow on the face of the Earth which moves across it as the Moon and Sun change their relative positions. Never is the umbra sufficiently large to engulf the whole Earth. Conse-

Figure 38. (*a*) Lunar eclipse. (*b*) Total solar eclipse. (*c*) Annular solar eclipse.

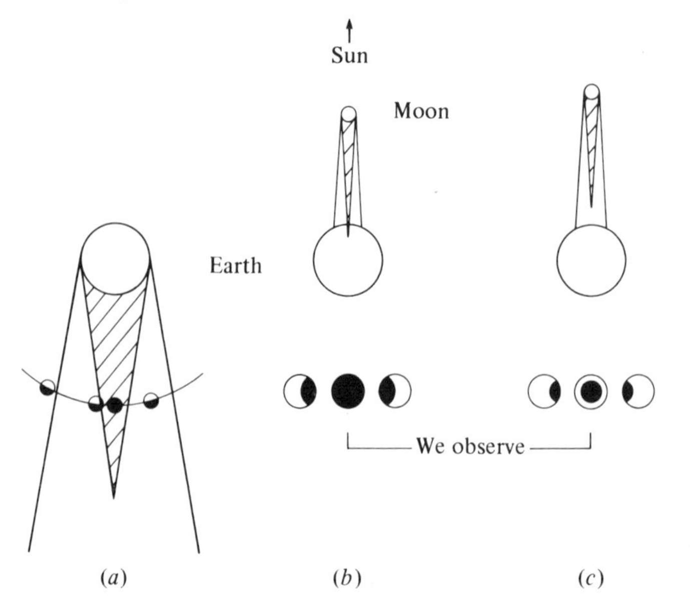

quently any total eclipse can only be seen along a narrow strip of the Earth's surface.

Sometimes, however, the umbra does not reach the Earth at all (Figure 38*c*). In this case an *annular eclipse* can occur with the Moon not quite obscuring the whole of the Sun's disc at maximum eclipse but leaving a ring of light round its edge.

68 The 'rules' of eclipses

Here is a summary of the most important 'rules' which appear to govern the occurrence of eclipses.

(*a*) A lunar eclipse can only occur at full Moon and a solar eclipse at new Moon. There is not an eclipse every month.

(*b*) At least two solar eclipses occur every year, and never more than five. There are a maximum of three lunar eclipses in a year. The highest total number of eclipses in a year, lunar and solar, is seven.

(*c*) Eclipses tend to go in pairs or threes: solar–lunar–solar. A lunar eclipse is always preceded or followed by a solar eclipse (two weeks in between them).

(*d*) The pattern of eclipses tends to recur in cycles of 18 years 11 days and 8 hours, the so-called 'Saros' cycle. The pattern is not repeated exactly.

(*e*) At the moment of greatest eclipse the Sun and Moon are either in opposition or conjunction. If the angle between the line of nodes and the Sun or Moon is greater than 12° 15′ a total lunar eclipse is not possible, while if it is less than 9° 30′ a lunar eclipse *must* occur. If the angle is more than 18° 31′ a solar eclipse cannot happen, while if it is less than 15° 31′ a solar eclipse *must* occur.

(*f*) In a lunar eclipse, the total phase can last for a maximum time of 1 hour 40 minutes, and the umbral phase, partial–total–partial, for a maximum time of 3 hours 40 minutes. The maximum time of total solar eclipse (at the equator) is 7 minutes 40 seconds and an annular eclipse can last at most for 12 minutes and 24 seconds.

69 Calculating a lunar eclipse

There are two questions to be asked before proceeding with the calculation of an eclipse. The first is 'Is an eclipse likely to occur?' If the answer is yes, the second is 'Will I be able to see it?' You may predict an eclipse of the Moon at a certain time, but if the Moon hasn't risen or has already set you won't be able to see it!

First, then, to spot the likely time of occurrence of a lunar eclipse. From rule (*a*) above we must have full Moon, that is, the angle $\lambda_m - \lambda_\odot = 180°$. From rule (*e*) the angle between the node and the Moon must be within 12° 15′ of 0° or 180° at that time. This is the angle $l'' - N'$ in the calculation of section 61.

For example, the results of the calculations for the Moon at 10h 41m UT on September 6th 1979 are:

$$\left.\begin{aligned} \lambda_\odot &= 163.24 \text{ degrees} \\ \lambda_m &= 343.08 \text{ degrees} \end{aligned}\right\} \quad \lambda_m - \lambda_\odot = 179.84 \text{ degrees}$$

$$\left.\begin{aligned} l'' &= 343.10 \text{ degrees} \\ N' &= 158.21 \text{ degrees} \end{aligned}\right\} \quad l'' - N' = 184.89 \text{ degrees}$$

We see that $\lambda_m - \lambda_\odot$ is very nearly 180° so that we are almost at the point of full Moon, and that $l'' - N' - 180 = 4.89$ degrees, well within the limit of 12° 15′. There was indeed a total lunar eclipse that day.

Secondly, can you see the eclipse? The answer obviously depends on your position on the Earth. We calculated in section 66 the times of moonrise and moonset on the day of the eclipse for an observer on the Greenwich meridian at latitude 52° N. We found that the Moon did not rise until 18h 38m UT while the eclipse was in progress at 10h 41m UT. Our observer did not therefore see the eclipse.

To calculate the circumstances of an eclipse we need to know the Moon's position at some particular time near to full Moon, its hourly motions in longitude and latitude, its angular distance from the Sun (the angle $\lambda_m - \lambda_\odot$), its angular diameter and the angular radius of the Earth's shadow at the distance of the Moon's orbit. This last is given by the following simple formula with sufficient accuracy for our purposes:

S_p = radius of penumbra = $\pi + 0.27$ degrees,

S_u = radius of umbra = $\pi - 0.27$ degrees,

where π is the Moon's horizontal parallax (section 65). Let us now calculate the circumstances of the lunar eclipse on September 6th 1979.

First, we write down all the details:

1979 September 6th at 10h 00m ET:

$\lambda_m = 342.643$ degrees

$\beta_m = -0.398$ degrees

$\lambda_\odot = 163.213$ degrees

$\Delta\lambda = 0.610$ degrees/hour

$\Delta\beta = -0.050$ degrees/hour

$\pi = 1.006$ degrees

$S_p = \pi + 0.27 = 1.276$ degrees

$S_u = \pi - 0.27 = 0.736$ degrees

$\theta_m = 0.548$ degrees

Add to this the fact that the Sun is also moving in longitude at a rate of $360/(365.2422 \times 24)$ degrees/hour, that is at 0.041 degrees/hour.

Next, we calculate the precise point of opposition. This is the moment when the angle $\lambda_m - \lambda_\odot = 180°$. At 10h 00m ET the angle was $342.643 - 163.213 = 179.430$ degrees, just 0.57 degrees short of 180°. The Moon was moving at a rate of 0.610 degrees/hour and the Sun at 0.041 degrees/hour. Thus the Moon was catching up on the Sun at a rate of $0.610 - 0.041 = 0.569$ degrees/hour, so that it took $0.57/0.569$ hours to catch up by 0.57 degrees, that is almost exactly one hour. Opposition (in ecliptic coordinates) was therefore at 11h 00m ET.

Now we are in a position to construct the *eclipse diagram*, Figure 39. Draw a horizontal line to represent the plane

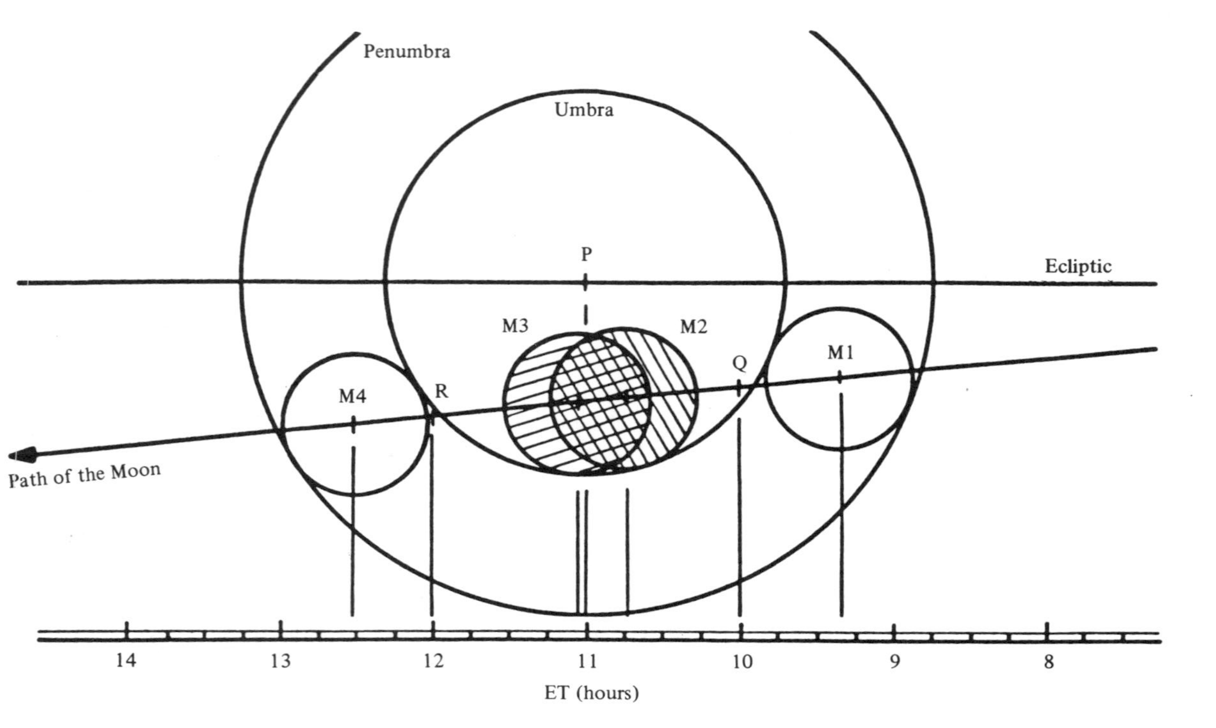

Figure 39. Lunar eclipse of September 6th 1979. Scale: 1 hour = 2 cm. GMT = ET − 50 seconds.

of the ecliptic. Below this, draw a parallel line and, choosing a suitable scale (say 1 hour = 2 cm), mark it in hours. Find the position on the scale corresponding to the moment of opposition and mark it (point P) on the ecliptic line. Next draw circles to scale centred on P to represent the umbra and the penumbra. Their radii should correspond with the angular radii of the shadows they represent. For example, we found above that $S_p = 1.276$ degrees. The difference between the hourly motions in longitude of the Sun and the Moon was 0.569 degrees/hour. Thus, on our scale of 1 hour = 2 cm, 0.569 degrees also equals 2 cm. Hence the scale factor between degrees and centimetres is 1 degree = 3.515 cm, and we should draw a circle of radius $1.276 \times 3.515 = 4.485$ cm to represent the penumbra.

Now we are ready to plot the path of the Moon through the Earth's shadow. First, we mark the position, Q, of the Moon at 10h 00m ET for which we have calculated its ecliptic coordinates. This position lies 0.398 degrees below the ecliptic, corresponding to $0.398 \times 3.515 = 1.40$ cm on the scale of our diagram. Then we calculate the Moon's position at some other time, say two hours later at 12h 00m ET. Mark this point R. Its position two hours later is

$$\beta = \beta_0 + \Delta\beta t = -0.398 + (-0.050 \times 2) = -0.498 \text{ degrees.}$$

This corresponds to 1.75 cm on the scale of our diagram. Joining Q and R with a straight line gives the path of the Moon.

Now we simply have to draw circles centred on the line RQ to represent the Moon at any point. The radii of these circles must correspond with the calculated angular radius of the Moon, $\theta/2$. In this case $\theta/2 = 0.274$ degrees or 0.96 cm on the scale of the diagram. We mark four such positions: M1 represents the point where the Moon enters the umbra, M2 where the total phase begins, M3 where it ends, and M4 where the Moon leaves the umbra. The corresponding times may be found from the scale below. They are:

Calculated	Astronomical Ephemeris
M1 at 09h 21m	09h 19m
M2 at 10h 44m	10h 32m
M3 at 11h 03m	11h 18m
M4 at 12h 31m	12h 31m

We see that, despite the simplicity of the method, we have calculated the times of the beginning and end of the umbral phase to within a couple of minutes. The errors in M2 and M3 may well be due to the thickness of the pencil line, for the lower limb of the Moon just grazes the lower edge of the umbra in this case.

70 Calculating a solar eclipse

A solar eclipse is rather more difficult to calculate than a lunar eclipse. If you look up a solar eclipse in the *Astronomical Ephemeris* you will find a map of the world showing the path and duration of the eclipse at each point; we shall not attempt such detail here. Our simple calculations will be made for just one location but will give a good guide of what to expect.

Once again we need to answer the question 'Is an eclipse likely?' Rule (*a*) in section 68 tells us that we have to be at new Moon, that is the angle $\lambda_m - \lambda_\odot$ equals 0° (or 360°). Rule (*e*) tells us that the angle between the Sun or the Moon and a node must be within 18° 31′ of 0° or 180° at that time; this is the angle $\lambda_\odot - N'$ or $l'' - N'$.

A solar eclipse was observed on February 26th 1979. We shall illustrate the method by working out the circumstances of that eclipse, as observed by someone on longitude 100° W and at latitude 50° N.

First, we must work through the calculations of sections 61, 62 and 65. The results for 16h 00m 50s ET on February 26th 1979 are:

$$\left.\begin{array}{l}\lambda_\odot = 337.448 \text{ degrees}\\ \lambda_m = 337.011 \text{ degrees}\end{array}\right\} \quad \lambda_m - \lambda_\odot = -0.437 \text{ degrees}$$

$\beta_m = 0.992$ degrees

$\Delta\lambda = 0.607$ degrees/hour

$\Delta\beta = -0.049$ degrees/hour

$\theta = 0.546$ degrees

$\rho'_m = 0.949$

$$\left.\begin{array}{l}l'' = 336.967 \text{ degrees}\\ N' = 168.097 \text{ degrees}\end{array}\right\} \quad l'' - N' = 168.87 \text{ degrees}$$

hourly motion of the Sun $= 0.041$ degrees/hour

We now have to take account of geocentric parallax. The coordinates λ_m and β_m that we have just calculated are those that would be observed at the centre of the Earth. We are observing from the surface of the Earth at position 100° W, 50° N, and we see slightly different ecliptic coordinates which can be calculated as follows:

Method	*Example*
1. Transform λ_m and β_m to equatorial coordinates (§ 27).	$\alpha_m =$ 22.557 500 hours $\delta_m =$ −8.016 944 degrees
2. Find the apparent right ascension and declination after allowing for parallax (§ 36).	$\alpha'_m =$ 22.587 070 degrees $\delta'_m =$ −8.844 591 degrees
3. Convert α'_m and δ'_m to λ'_m and β'_m using the method of § 28.	$\lambda'_m =$ **337.113 312 degrees** $\beta'_m =$ **0.060 038 degrees**

Next we calculate the precise moment of conjunction in ecliptic coordinates when $\lambda'_m - \lambda_\odot = 0°$. Strictly, we ought to apply the correction for parallax to the Sun's coordinates as well, but we shall ignore this small correction here. At

16h 00m 50s ET, $\lambda'_m - \lambda_\odot = -0.335$ degrees so that the Moon had still a little distance to catch up with the Sun. Its speed in longitude was $\Delta\lambda = 0.607$ degrees/hour so it was gaining on the Sun at $0.607 - 0.041 = 0.566$ degrees/hour. The difference of 0.335 degrees was made up in 0.335/0.566 hours = 0.627 hours = 38 minutes. Conjunction therefore occurred at 16h 39m ET. At this moment the Sun's longitude was $337.448 + (0.627 \times 0.041) = 337.474$ degrees. The value of $\lambda_\odot - N' - 180$ was therefore -10.623 degrees, well within the limit set by rule (*e*). We deduce that the eclipse *must* have occurred (as we know it did!).

Now we are in a position to construct the eclipse diagram (Figure 40). We proceed exactly as we did for the lunar eclipse, drawing two horizontal lines, one to represent the ecliptic and the lower one to represent time. Choosing a suitable scale (say 2 cm = 1 hour) we mark off the lower line in hours such that the time of conjunction is roughly in the middle of the diagram. Next we find point P on the ecliptic corresponding to conjunction and we draw a circle centred on P of the correct radius to represent the Sun. In this calculation, we assume that the angular radius of the Sun is 0.27 degrees. The relative motion between Sun and Moon in this case is 0.566 degrees/hour. Thus 2 cm on our scale, which represents 1 hour, also represents 0.566 degrees. Hence 1 degree = 3.534 cm. The Sun's radius converts to $0.27 \times 3.534 =$ 0.954 cm on the scale of the diagram.

Next we plot the position of the Moon which we have calculated, using the corrected value β'_m. We find the point corresponding to 16h 00m 50s ET and $\beta'_m = 0.060$ degrees (= 0.212 cm), and mark it Q. Then we find the Moon's position, say, two hours later using the hourly motion in latitude:

$$\beta_m = 0.060 - (0.049 \times 2) = -0.038 \text{ degrees} = -0.13 \text{ cm}.$$

Mark this point R. Joining Q and R by a straight line gives the path of the Moon relative to the Sun. We need only mark off

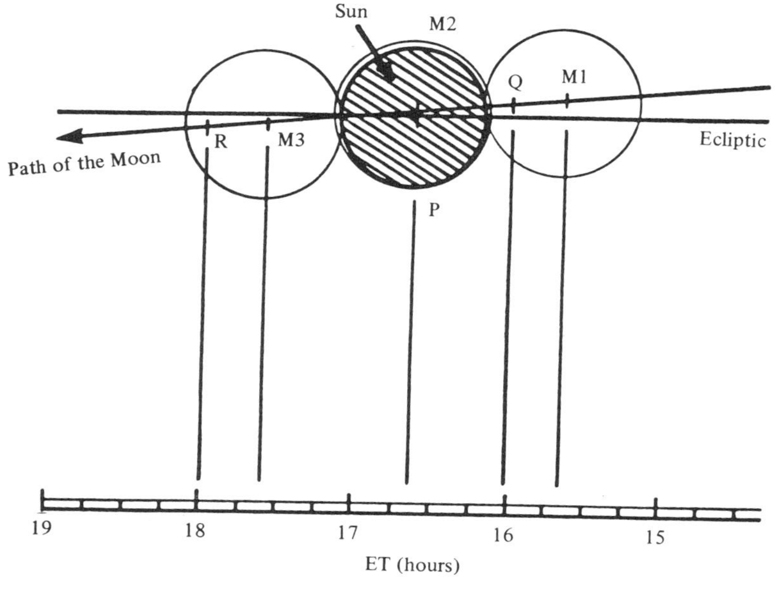

Figure 40. Solar eclipse of February 26th 1979 as observed from longitude 100° W, latitude 50° N.

circles centred on the line of the correct radius ($\theta/2$) to represent the Moon to determine the aspect of the eclipse at any time. In Figure 40 we have marked three positions, M1, M2 and M3, corresponding to the start, middle and end of the eclipse. The circles are of radius $(0.546/2)\times 3.534 = 0.97$ cm. Here are the results:

Calculated	Astronomical Ephemeris
M1 at 15h 39m	15h 37m
M2 at 16h 39m	16h 50m
M3 at 17h 35m	17h 52m

You will notice from Figure 40 that the eclipse was total. Comparison of our calculated times with those deduced from the maps and tables of the *Astronomical Ephemeris* shows that we are within about a quarter of an hour of the correct results. Even our comparatively simple method allows us to make quite accurate predictions of what is surely the heavens' most awe-inspiring phenomenon.

71 The Astronomical Calendar

It is often useful to have a chart which shows, at a glance, the relative configurations of the Sun, Moon and planets and the likely times of occurrence of eclipses. The Astronomical Calendar is just such a chart, displaying the right ascension of each heavenly body for every day in the year; the chart for 1980 is drawn in Figure 41.

It is convenient (though not essential) to construct the chart on graph paper. Mark the vertical axis in days (1 to 365 or 366) on a scale to make best use of the paper, and the horizontal axis in hours (0 to 24) such that time increases towards the left;

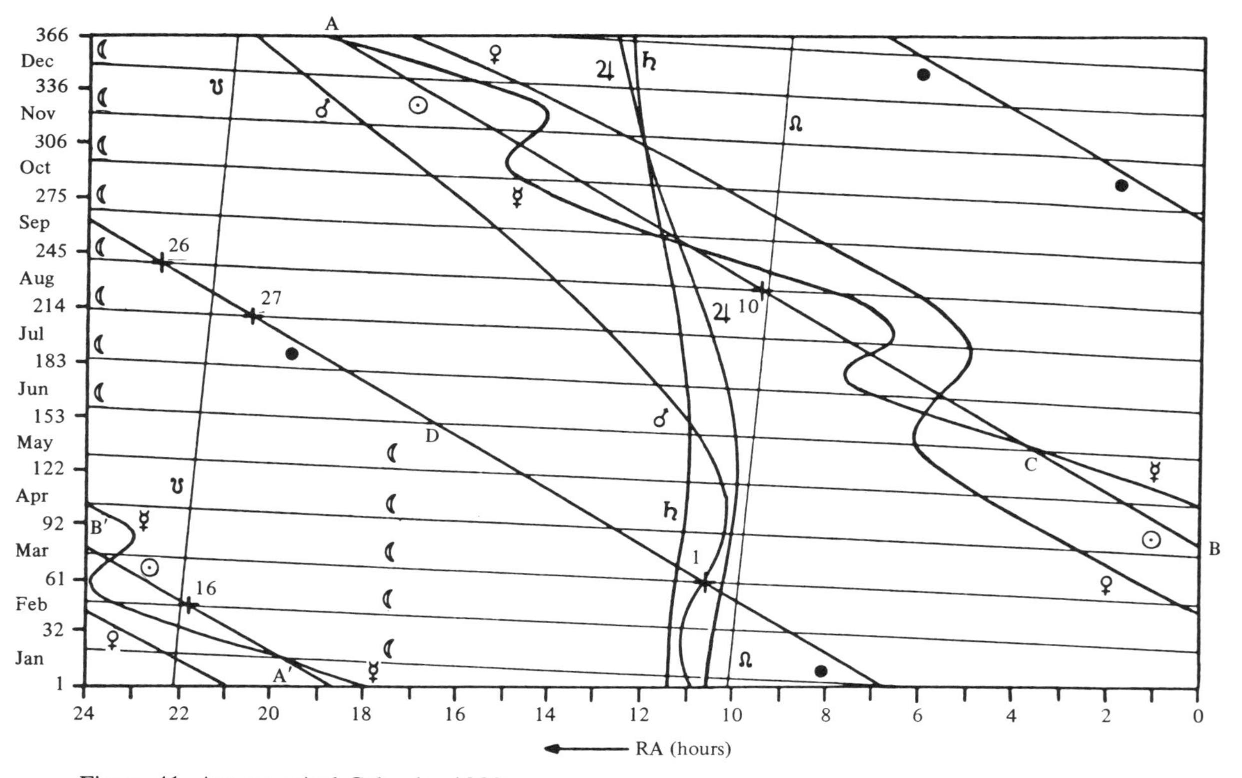

Figure 41. Astronomical Calendar 1980.

this is a convention often adopted by astronomers as the chart then more nearly represents the relative positions of the bodies in the sky as seen from the Earth. Lines representing the times of the Sun and midnight may now be drawn. To do so it is necessary only to calculate the right ascension of the Sun on two days of the year several months apart by the method of section 42, and to join the points by a straight line. Where the line goes off the edge of the chart (as at A and B in Figure 41) it should be continued from the points exactly opposite (A′ and B′). The resulting lines should slope down towards the right; mark them with the symbol ⊙ to represent the Sun. The tracks of midnight, marked by the symbol ●, are parallel to those of the Sun but displaced by 12 hours on the RA (right ascension) scale.

Next, the track of the Moon should be marked in. Again, this can be done by calculating the Moon's right ascension on two days (a week or so apart) every month using the method of section 61 and joining the points by straight lines. However, the calculations are lengthy and somewhat tedious unless you have a programmable calculator, so you may like to cheat a bit. The position of the Moon is given for every hour of the year in the *Astronomical Ephemeris* which you can consult in your local library, but it can also be deduced from the information given in most diaries, the dates of new Moon and full Moon. We know that when the Moon is full it is in opposition to the Sun and, conversely, it is in conjunction with the Sun at new Moon. We have already marked the tracks of conjunction (⊙) and opposition (●) so that we can easily plot the Moon's position from the dates of new Moon and full Moon. For example, my diary indicated that new and full Moons occurred on May 14th and 29th 1980 respectively. The Moon's right ascension on the 14th was therefore the same as that of the Sun and on the 29th it was the same as that of midnight. Join these two points (C and D in Figure 41) with a straight line to mark the track of the Moon.

Next we must mark in the tracks of the Moon's ascending node (☊) and descending node (☋). The mean longitude of the former is given by the value of N in section 61, and of the latter by $N+180$. Find these values on two days separated by six months or so and convert to right ascension by the method of section 27 (setting $\beta = 0$). Join each pair of points by a straight line.

We are now in a position to make predictions about eclipses. As explained in section 68, an eclipse can only occur when the Moon is near one of its nodes at full Moon (lunar eclipse) or new Moon (solar eclipse). We must therefore find points on the chart where the tracks of the Moon, Sun or midnight, and either node pass close to one another. In Figure 41 these points are marked '+' together with the dates on which the eclipses occurred as follows:

February 16:	Total eclipse of the Sun
March 1:	Penumbral eclipse of the Moon
July 27:	Penumbral eclipse of the Moon
August 10:	Annular eclipse of the Sun
August 26:	Penumbral eclipse of the Moon

To complete the chart, we can mark on the tracks of the major planets: Mercury (☿), Venus (♀), Mars (♂), Jupiter (♃), Saturn (♄), Uranus (♅), Neptune (♆) and Pluto (♇). Their positions can be calculated by the method given in section 50.

Glossary of terms

age of Moon: the angle between the Sun and the Moon measured at the Earth.

altitude: the angle up from the horizon.

annual equation: a correction of the Moon's orbital motion due to the variation of the Sun–Earth distance as the Earth travels in its own ellipse about the Sun.

anomaly: the angle at the focus or the centre of an orbital ellipse between the major axis and the orbiting body or its projection. The *eccentric anomaly*, E, is defined in Figure 19, section 43, while the *mean anomaly*, M, and *true anomaly*, ν, are defined in Figure 18, section 42.

apastron: the point in an orbit about a star that is furthest from the star.

aphelion: the point on an orbit about the Sun most distant from the Sun.

apogee: the point on an orbit about the Earth most distant from the Earth.

Astronomical Ephemeris: a collection of tables predicting the positions and circumstances of astronomical phenomena.

astronomical latitude: the angle between the astronomical zenith and the equator.

astronomical unit: the length of the semi-major axis of the Earth's orbit about the Sun.

atmospheric refraction: the apparent shift in the position of a celestial object due to the bending of light rays by the atmosphere.

azimuth: the angle round from the north point measured on the horizon in the sense NESW.

calendar: system of accounting the days in the year. The *Julian calendar*, introduced by Julius Caesar, divides the year into 365 days except for every fourth year which has 366. The *Gregorian calendar*, introduced by Pope Gregory in 1582 and accepted in

England in 1752, is the one in use today. It reduced the errors in the Julian calendar by removing three days every four centuries; if the year ends in two noughts it is only a leap year if it is divisible by 400.

celestial sphere: an imaginary sphere, usually centred on the Earth, of arbitrary large radius on the surface of which the stars can be considered to be fixed.

circumpolar stars: stars whose angular distances from the north or south pole are sufficiently small that they never dip below the horizon.

comet: a diffuse member of the Solar System, usually with a highly elongated orbit, which becomes visible near the Sun. It has a bright head and a diffuse tail of variable length which always points away from the Sun.

companion star: the darker of the pair of stars in a visual-binary system.

conjunction: the moment when two celestial bodies occupy the same position in the sky or share a common coordinate when viewed from a particular place. Thus *heliocentric conjunction*, and *conjunction in right ascension.*

coordinate systems: frames of reference by means of which the position of any point can be uniquely specified. In astronomy, the systems take their names from the fundamental planes on which they are based. Thus the *ecliptic coordinate system* measures longitude round from the first point of Aries, ♈, in the plane of the ecliptic and latitude northwards from it. The *equatorial coordinate system* measures right ascension round from ♈ in the plane of the Earth's equator, and declination northwards from it. In the *horizon coordinate system*, the azimuth is measured round from the north point in the sense NESW and the altitude is the angle up from the horizon. The *galactic coordinate system* specifies position by longitude measured in the galactic plane round from the direction of the galactic centre and by latitude measured perpendicular to the plane. *Heliographic coordinates* enable the position of an object on the surface of the Sun to be specified with respect to the solar equator and a fundamental meridian assumed to rotate at a uniform rate.

culmination: the moment at which a celestial body crosses the observer's meridian. Circumpolar stars cross the meridian above the horizon twice in one day, giving *upper culmination* and *lower culmination.*

day: the interval between two successive transits across the observer's meridian of a fixed star (*sidereal day*), of the Sun (*solar day*), or of a fictitious body called the mean Sun which moves at a uniform rate along the equator (*mean solar day*).
declination: in the equatorial coordinate system, the angle measured perpendicular to the equator (north positive, south negative).

earthshine: light reflected from the Earth which sometimes illuminates the dark portion of the Moon's disc, making it visible.
eccentricity: a measure of the degree of elongation of an ellipse, equal to the ratio

the distance of the focus from the centre: the semi-major axis.

eclipse: the passage of the Moon through the Earth's shadow (*lunar eclipse*) or parts of the Earth through the Moon's shadow (*solar eclipse*). If, at the moment of greatest eclipse, the Moon or Sun is only partly obscured it is a *partial eclipse*; if completely obscured it is a *total eclipse*. If during a solar eclipse the Moon obscures the central part of the Sun's disc but leaves an unobscured ring around its edge, then it is an *annular eclipse*.
ecliptic: the plane containing the orbit of the Earth about the Sun.
ellipse: a type of regular closed curve, oval in shape, of which a circle is a special case. It is traced by a point moving in such a manner that it keeps constant the sum of its distances from two fixed points, each of which is called a *focus* of the ellipse. The longest diameter of the ellipse, which goes through both foci and the centre, is called the *major axis*, the portion from the centre to the curve in either direction being called the *semi-major axis*.
epoch: a particular moment specified as the reference point from which time is measured.
equation of the centre: a relation between the true and mean anomalies which is an approximation to Kepler's equation. In its simplest form it is

$$\nu = M + 2e \sin M$$

where ν and M are expressed in radians.
equation of time: the difference between the mean solar time and the real solar time.
equator: the plane through the centre of the Earth which is perpendicular to the spin axis.
equinox: the moment at which the Sun crosses the celestial equator. This occurs at about March 21st when its right ascension is zero

(the *vernal equinox*) and about September 22nd when its right ascension is 12h (the *autumnal equinox*). The positions of the equinoxes on the celestial sphere lie along the line of the intersection of the planes of the equator and the ecliptic.

evection: a correction to the Moon's orbital motion taking account of slight variations in the apparent value of the eccentricity of its orbit.

extinction: the attenuation and colouring of light as it travels through a medium; in particular, *atmospheric extinction.*

figure of the Earth: the true shape of the Earth. It is often approximated by a *spheroid of revolution*, a geometrical shape in which any cross-section parallel to the equator is a circle, while any cross-section through the north–south axis is an ellipse with the minor axis coincident with the diameter joining the north and south poles.

first point of Aries: the position on the celestial sphere of the vernal equinox.

focus of an ellipse: see *ellipse.*

geocentric coordinates: coordinates measured with respect to the centre of the Earth. Hence the *geocentric latitude* is the angle between the equator and a point on the surface of the Earth, as measured at the centre of the Earth.

geocentric parallax: the angle subtended at a heavenly body by the centre of the Earth and the point of observation on the Earth's surface.

geostationary satellite: a body orbiting the Earth in the plane of the equator in such a direction and at such a height that its orbital period equals one day so that it keeps constant position with respect to the Earth's surface.

gravity: the mutual force of attraction between any two bodies which is proportional to the product of their masses and inversely proportional to the square of their separation.

great circle: any circle drawn on the surface of a sphere whose centre is the same as that of the sphere.

Greenwich meridian: that half of the great circle on the surface of the Earth passing through the north and south poles and through the reference point in Greenwich, England. It is taken as the line of longitude 0°.

horizontal parallax: the geocentric parallax when the celestial body is on the observer's horizon; hence *equational horizontal parallax* when the observer is also on the equator.
hour-angle: the difference between the local sidereal time and the right ascension.

inclination of orbit: the angle between the plane of the orbit and the plane of the ecliptic.
inner planet: a planet whose semi-major axis is less than that of the Earth; that is, the planets Mercury and Venus.

Julian day number: the number of Julian days which have elapsed since the fundamental epoch Greenwich mean noon of January 1st 4713 B.C. For January 1st 1975, 12h UT, this was 2 442 414. See also *modified Julian date*.

Kepler's equation: the relation between the mean and eccentric anomalies

$$E - e \sin E = M,$$

where the angles are expressed in radians.

latitude: the coordinate expressing the angle (north positive, south negative) perpendicular to a fundamental plane, hence *ecliptic latitude* and *galactic latitude*. On the Earth, the *geographical latitude* is measured with respect to the equator. The ecliptic latitude can either be measured at the Earth (*geocentric*) or at the Sun (*heliocentric*).
longitude: the coordinate expressing the angle round from a fixed direction measured in a fundamental plane, hence *ecliptic longitude* and *galactic longitude*. On the Earth, the *geographical longitude* is measured at the equator. The ecliptic longitude can either be measured at the Earth (*geocentric*) or at the Sun (*heliocentric*).
luni-solar precession: the slow retrograde motion of the first point of Aries along the equator due to the combined effects of the Sun and the Moon on the slightly non-spherical Earth.

magnitude: the unit defined on a logarithmic scale which measures the brightness of a celestial object.
mean Sun: a fictitious heavenly body which moves at a uniform rate along the equator making one complete circuit in the same time (one year) as the real Sun takes to make a complete circuit.

meridian: that half of a great circle which is terminated at the north and south poles. On the Earth a meridian is a line of longitude. On the celestial sphere, the meridian which passes through the zenith is called the *observer's meridian.*

modified Julian date: the number of Julian days elapsed since 1858 November 17.0.

month: the period taken by the Moon to make one complete circuit of its orbit from reference point to reference point. The *draconic month* or *nodal month* takes the ascending node as the reference and is equal to 27.2122 mean solar days. The *sidereal month* is reckoned against the background of stars and is equal to 27.3217 mean solar days. The Sun is used as the reference for the *synodic month* of 29.5306 mean solar days, and the perigee in the *anomalistic month* of 27.5546 mean solar days.

node: the point on the celestial sphere where the great circle representing the orbit cuts the great circle representing the plane of the ecliptic. The point where the orbiting body is moving from below (south of) to above the ecliptic is called the *ascending node*; the other is the *descending node.*

noon: the instant at which the Sun crosses the observer's meridian.

north celestial pole: the point at which the projection of the Earth's rotation axis through the north pole intersects the celestial sphere.

nutation: a small periodic wobbling motion of the Earth's rotation axis.

obliquity of the ecliptic: the angle at which the plane of the ecliptic is inclined to the plane of the equator.

opposition: the moment when two celestial bodies occupy opposite positions in the sky, or have longitudes different by 180°, when viewed at a particular place.

orbit: the path through space taken by a body gravitationally bound to another body.

orbital elements: the quantities which need to be known to specify an orbit uniquely.

osculating elements: the elements describing the elliptical orbit followed by a body if all perturbing influences vanish. Since perturbations disturb the true orbit of any member of the Solar System, the osculating elements are constantly changing.

outer planets: those planets having semi-major axes larger than that of the Earth. The major outer planets are Mars, Jupiter, Saturn, Uranus, Neptune and Pluto.

parabolic orbit: an orbit in which the velocity at any point is equal to the escape velocity.
parallax: the amount by which the apparent position of a celestial object shifts as the point of observation is changed.
penumbra: the outer portion of a shadow where the light is only partially cut off.
periastron: the point in an orbit about a star that is nearest to the star.
perigee: the point on an orbit about the Earth which is nearest the Earth.
perihelion: the point of closest approach to the Sun in an orbit about the Sun.
period of orbit: the time taken by the orbiting body to make one complete circuit.
perturbations: deviations from true elliptical motion caused by the gravitational fields of other members of the Solar System.
phase: (i) *of Moon or planet*: the area of the disc which is illuminated. When the dark side of the Moon faces the Earth, the phase is zero and it is *new Moon.* At the *first quarter* and the *third quarter*, the phase is equal to a half and the Moon is *in quadrature. Full Moon* has a phase equal to one. Whenever the phase is greater than a half, the Moon is described as *gibbous.*

(ii) *of an eclipse*: the stage of a lunar or solar eclipse during which the eclipsed body is partly obscured (*partial phase*) or totally obscured (*total phase*). During a lunar eclipse, the Moon is in the penumbra of the Earth's shadow during the *penumbral phase* and partially or totally in the umbra during the *umbral phase.* The partial and total phases occur during the umbral phase.
planet: a solid body in closed orbit about a star. In our own Solar System, the major planets are (in order of increasing distance from the Sun) Mercury, Venus, Earth, Mars, Jupiter, Saturn, Uranus, Neptune and Pluto.
polar distance: the angle on the celestial sphere from the celestial pole.
pole: the point on a sphere which is perpendicular to a given plane. Hence *pole of the ecliptic* and *pole of the equator* (each has two poles called north and south poles for short).
position-angle: a celestial angle measured from 0° to 360° eastwards from the north.
precession: see *luni-solar precession.*
primary star: the brighter of the pair of stars in a visual-binary system.
prograde motion: motion in the same sense as that of all the planets

about the Sun. When looking down on the Solar System from the north celestial pole, prograde motion is counter-clockwise.

radius vector: the line joining the principal focus to the position of the orbiting body on its orbital ellipse.

reflectivity of planet: a measure of a planet's ability to reflect sunlight; a factor affecting its apparent brightness.

refraction: see *atmospheric refraction.*

retrograde motion: motion in the opposite sense to that of all the planets about the Sun. When looking down on the Solar System from the north celestial pole, retrograde motion is clockwise.

right ascension: in the equatorial coordinate system the angle measured round from the first point of Aries in the plane of the equator, in the sense SENW.

rising: the moment when a celestial body crosses the horizon on the way up.

Saros cycle: the period of 18 years 11 days and 8 hours after which the pattern of lunar and solar eclipses tends to repeat.

semi-major axis: see *ellipse.*

setting: the moment when a celestial body crosses the horizon on the way down.

solar elongation: the angle between the lines of sight to the Sun and to the celestial body in question.

Solar System: the Sun and all the bodies, planets, comets and asteroids in closed orbits about it.

synodic period: the time between successive conjunctions in longitude.

terminator: the line marking the boundary between the dark and sunlit hemispheres of a member of the Solar System.

time: (i) *solar time*: time measured with respect to the motion of the Sun or a fictitious body called the mean Sun (*mean solar time*). *Greenwich mean time*, or *universal time* as it is also known, is the mean solar time as measured on the Greenwich meridian. Recent measurements with highly accurate atomic clocks have shown the rotation period of the Earth to be irregular and *ephemeris time* has been introduced which advances at a uniform rate irrespective of the Earth's motion. Ephemeris time and universal time are almost equal. *British summer time* is one hour ahead of Greenwich mean time and is an example of *daylight saving time* in which the time is adjusted to make the working day fit more conveniently into daylight hours.

(ii) *sidereal time*: time measured with respect to the motion of the stars. The *local sidereal time* at any place is equal to the hour-angle of the first point of Aries which on the Greenwich meridian is called *Greenwich sidereal time.* The difference between *apparent sidereal time* and *mean sidereal time* takes account of a small periodic wobbling motion of the Earth's rotation axis called nutation and may amount numerically to 1.2 seconds. The former is associated with the motion of the true ♈ while the latter is regulated by a fictitious mean ♈ in which the nutation has been averaged out.

time zone: longitudinal strip on the surface of the Earth in which the *zone time*, a whole number of hours before or after GMT, is adopted as the local civil time by national or international agreement.

transit: the moment at which a celestial body crosses the observer's meridian.

twilight: that period of semi-darkness after sunset or before sunrise during which the sun's zenith distance is more than 90° but less than some agreed figure. This figure is 108° for *astronomical twilight*, while for *civil twilight* it is 96°.

umbra: the inner portion of a shadow where the light is completely obscured.

variation: a correction to the Moon's orbital motion about the Earth which takes account of the changing solar gravitational field.

year: the interval between two successive passages of the Sun through a reference point. A particular point among the stars is used as reference in the *sidereal year*, equal to 365.2564 mean solar days. The *tropical year*, 365.2422 mean solar days, uses the first point of Aries as its reference. When no qualifying adjective is used with the word 'year', it is usually the tropical year that is meant. Perturbations to the Earth's orbit by the other planets cause small changes in the Earth's orbital elements. The *anomalistic year*, 365.2596 mean solar days, is the interval between two successive passages of the Sun through perigee. The *Besselian year* is the same as the tropical year but defines the moment at which it begins as the instant when the right ascension of the mean Sun is exactly 280° or 18h 40m; this instant falls very near the beginning of the civil year. Strictly speaking, it is the Besselian year that we ought to use in astronomical calculations.

zenith: the point directly overhead at the observer. The *zenith angle* or *zenith distance* of a star is the angle between the star and the zenith.

zone correction: the number of hours that needs to be added to or subtracted from GMT to get the *zone time*.

Symbols and abbreviations

☉	Sun	♄	Saturn
●	midnight	♅	Uranus
☾	Moon	♆	Neptune
☿	Mercury	♇	Pluto
♀	Venus	♈	first point of Aries
⊕	Earth	☊	ascending node
♂	Mars	☋	descending node
♃	Jupiter		

α	right ascension
β	geocentric ecliptic latitude
δ	declination
Δ	difference; error
ε	elongation; obliquity of the ecliptic; longitude of planet at epoch
ε_g	geocentric longitude of Sun at epoch
ζ	apparent zenith angle
θ	angular diameter; displacement
λ	geocentric ecliptic longitude
ν	true anomaly
π	parallax; constant = 3.141 592 654
ϖ	heliocentric longitude of perihelion
ϖ_g	geocentric longitude of Sun's perigee
ρ	distance
τ	light-travel time
ϕ	geographical latitude
ϕ'	geocentric latitude
χ	position-angle
ψ	heliocentric ecliptic latitude; angle at the horizon
☊	longitude of ascending node
ω	argument of perihelion

$\Delta\lambda$	Moon's hourly motion in ecliptic longitude
$\Delta\beta$	Moon's hourly motion in ecliptic latitude
ΔA	correction to azimuth
ΔT	ET–UT
A	azimuth; planet's brightness factor
A_e	annual equation
A_3	third correction to Moon's mean anomaly
A_4	fourth correction to Moon's mean anomaly
AU	astronomical unit
a	altitude; semi-major axis
B	heliographic latitude
BST	British summer time
b	galactic latitude
CRN	Carrington Rotation Number
DEC	declination
D	age of Moon or planet; number of days since an epoch
d	number of days; angle
E	eccentric anomaly; value of equation of time
E	east point of horizon
E_c	correction applied in the equation of the centre
ET	ephemeris time
E_v	evection
e	eccentricity
F	phase
GMT	Greenwich mean time
GST	Greenwich sidereal time
H	hour-angle
I	inclination of Sun's equator
i	inclination
JD	Julian days
L	heliocentric longitude of Earth or heliographic longitude
LST	local sidereal time
l	galactic longitude; Moon's orbital longitude; heliocentric longitude of planet
M	mean anomaly
MJD	modified Julian date
m	magnitude; precession constant
N	north point of horizon; longitude of ascending node
NCP	north celestial pole
NP	north pole
n	precession constant

P equatorial horizontal parallax; angle
p parallax
q perihelion distance
R refraction angle; distance of Earth from Sun
RA right ascension
r radius vector
r_0 semi-major axis of orbit
S south point on horizon
SCP south celestial pole
SP south pole
S_p radius of Earth's penumbra
ST sidereal time
S_u radius of Earth's umbra
T_p period of orbit
t, t_0 epoch
UT universal time
V variation
W west point on horizon
Y years
z real zenith angle

Index